Lecture Notes in Artificial Intelligence   10997

Subseries of Lecture Notes in Computer Science

More information about this series at http://www.springer.com/series/1244

Dietmar Seipel · Michael Hanus
Salvador Abreu (Eds.)

Declarative Programming and Knowledge Management

Conference on Declarative Programming, DECLARE 2017
Unifying INAP, WFLP, and WLP
Würzburg, Germany, September 19–22, 2017
Revised Selected Papers

 Springer

Editors
Dietmar Seipel
Universität Würzburg
Wuerzburg
Germany

Salvador Abreu (iD)
Universidade de Èvora
Evora
Portugal

Michael Hanus (iD)
Christian-Albrechts-Universität zu Kiel
Kiel
Germany

ISSN 0302-9743   ISSN 1611-3349   (electronic)
Lecture Notes in Artificial Intelligence
ISBN 978-3-030-00800-0   ISBN 978-3-030-00801-7   (eBook)
https://doi.org/10.1007/978-3-030-00801-7

Library of Congress Control Number: 2018954670

LNCS Sublibrary: SL7 – Artificial Intelligence

This Springer imprint is published by the registered company Springer Nature Switzerland AG
The registered company address is: Gewerbestrasse 11, 6330 Cham, Switzerland

Preface

This volume contains a selection of the papers presented at the International Conference on Declarative Programming Declare 2017. The joint conference was held in Würzburg, Germany, during September 19–22, 2017. It consisted of the 21st International Conference on Applications of Declarative Programming and Knowledge Management (INAP), the 31st Workshop on Logic Programming (WLP), and the 25th Workshop on Functional and (Constraint) Logic Programming (WFLP), and it was accompanied by a one-week summer school on Advanced Concepts for Databases and Logic Programming for students and PhD students.

Declarative programming is an advanced paradigm for modeling and solving complex problems, which has attracted increased attention over the last decades, e.g., in the domains of data and knowledge engineering, databases, artificial intelligence, natural language processing, modeling and processing combinatorial problems, and for establishing knowledge-based systems for the web. The conference Declare 2017 aimed to promote the cross–fertilizing exchange of ideas and experiences among researches and students from the different communities interested in the foundations, applications, and combinations of high-level, declarative programming and related areas.

The INAP conferences provide a forum for intensive discussions of applications of important technologies around logic programming, constraint problem solving, and closely related advanced software. They comprehensively cover the impact of programmable logic solvers in the Internet society, its underlying technologies, and leading edge applications in industry, commerce, government, and societal services. Previous INAP conferences have been held in Japan, Germany, Portugal, and Austria. The Workshops on Logic Programming (WLP) are the annual meeting of the German Society for Logic Programming (GLP e.V.). They bring together international researchers interested in logic programming, constraint programming, and related areas like databases and artificial intelligence. Previous WLP workshops have been held in Germany, Austria, Switzerland, and Egypt. The International Workshop on Functional and Logic Programming (WFLP) brings together researchers interested in functional programming, logic programming, as well as the integration of these paradigms. Previous WFLP editions have been held in Germany, France, Spain, Italy, Estonia, Brazil, Denmark, and Japan. The topics of the papers of this year's joint conference Declare concentrated on three currently important fields: constraint programming and solving, functional and logic programming, and declarative programming.

The declarative programming paradigm expresses the logic of a computation in an abstract way. Thus, the semantics of a declarative language becomes easier to grasp for domain experts. Declarative programming offers many advantages for data and knowledge engineering, such as, e.g., security, safety, and shorter development time. During the last couple of years, a lot of research has been conducted on the usage of declarative systems in areas like answer set programming, reasoning, meta-programming, and deductive databases. Reasoning about knowledge wrapped in rules, databases, or the

Semantic Web allows to explore interesting hidden knowledge. Declarative techniques for the transformation, deduction, induction, visualization, or querying of knowledge have the advantage of high transparency and better maintainability compared to procedural approaches.

Many problems which occur in large industrial tasks are intractable, invalidating their solution by exact or even many approximate constructive algorithms. One approach which has made substantial progress over the last few years is constraint programming. Its declarative nature offers significant advantages, from a software engineering standpoint and in the specification, implementation, and maintenance phases. Several interesting aspects are in discussion: how can this paradigm be improved or combined with known, classical methods; how can real-world situations be modelled as constraint problems; what strategies may be pursued to solve a problem once it has been specified; or what is the experience of applications in really large industrial planning, simulation, and optimisation tasks?

Another area of active research is the use of declarative programming languages, in particular, functional and logic languages, to implement more reliable software systems. The closeness of these languages to logical models provides new methods to test and verify programs. Combining different programming paradigms is beneficial from a software engineering point of view. Therefore, the extension of the logic programming paradigm and its integration with other programming concepts are active research branches. The successful extension of logic programming with constraints has already been mentioned. The integration of logic programming with other programming paradigms has been mainly investigated for the case of functional programming, so that types, modules, higher-order operators, or lazy evaluation can also be used in logic-oriented computations.

The three events INAP, WLP, and WFLP were jointly organized by the University of Würzburg and the Society for Logic Programming (GLP e.V.). We would like to thank all authors who submitted papers and all conference participants for the fruitful discussions. We are grateful to the members of the Program Committee and the external referees for their timely expertise in carefully reviewing the papers. We would like to express our thanks to the German Federal Ministry of Education and Research (BMBF) for funding the summer school on Advanced Concepts for Databases and Logic Programming (under 01PL16019) and to the University of Würzburg for hosting the conference in the new Central Lecture Building Z6 and for providing the Tuscany Hall in the Baroque style Würzburg Residence Palace for a classical music concert in honor of Jack Minker, a pioneer in deductive databases and disjunctive logic programming and the longtime mentor of the first editor, who celebrated his 90th birthday in 2017.

July 2018

Dietmar Seipel
Michael Hanus
Salvador Abreu

Organization

Program Chair

Dietmar Seipel   University of Würzburg, Germany

Program Committee of INAP

Slim Abdennadher   German University in Cairo, Egypt
Salvador Abreu (Co-chair)   University of Évora, Portugal
Molham Aref   Logic Blox Inc, Atlanta, USA
Chitta Baral   Arizona State University, USA
Joachim Baumeister   University of Würzburg, Germany
Stefan Brass   University of Halle, Germany
François Bry   Ludwig-Maximilian University of Munich, Germany
Philippe Codognet   University Pierre-and-Marie Curie, France
Vitor Santos Costa   University of Porto, Portugal
Agostino Dovier   University of Udine, Italy
Thomas Eiter   Vienna University of Technology, Austria
Thom Frühwirth   University of Ulm, Germany
Parke Godfrey   York University, Canada
Gopal Gupta   University of Texas at Dallas, USA
Michael Hanus   Kiel University, Germany
Jorge Lobo   ICREA and Universitat Pompeu Fabra, Spain
Grzegorz J. Nalepa   AGH University, Poland
Vitor Nogueira   University of Évora, Portugal
Enrico Pontelli   New Mexico State University, USA
Dietmar Seipel (Chair)   University of Würzburg, Germany
Hans Tompits   Vienna University of Technology, Austria
Masanobu Umeda   Kyushu Institute of Technology, Japan

Program Committee of WLP/WFLP

Slim Abdennadher   German University in Cairo, Egypt
Sergio Antoy   Portland State University, USA
Olaf Chitil   University of Kent, UK
Jürgen Dix   Clausthal University of Technology, Germany
Moreno Falaschi   Università di Siena, Italy
Michael Hanus (Chair)   Kiel University, Germany
Sebastiaan Joosten   University of Innsbruck, Austria
Oleg Kiselyov   Tohoku University, Japan
Herbert Kuchen   University of Münster, Germany
Tom Schrijvers   Katholieke Universiteit Leuven, Belgium

Sibylle Schwarz	HTWK Leipzig, Germany
Dietmar Seipel	University of Würzburg, Germany
Martin Sulzmann	Karlsruhe University of Applied Sciences, Germany
Hans Tompits	Vienna University of Technology, Austria
German Vidal	Universitat Politècnica de València, Spain
Janis Voigtländer	University of Duisburg-Essen, Germany
Johannes Waldmann	HTWK Leipzig, Germany

Local Organization

Falco Nogatz	University of Würzburg, Germany
Dietmar Seipel	University of Würzburg, Germany

Additional Reviewers

Pedro Barahona
Zhuo Chen
Daniel Gall
Falco Nogatz
Nada Sharaf

Contents

Functional and Logic Programming

Constraints

Constraints

Constraint Solving on Hybrid Systems

Pedro Roque(✉) and Vasco Pedro

LISP, Universidade de Évora, Évora, Portugal
d11735@alunos.uevora.pt, vp@di.uevora.pt

Abstract. Applying parallelism to constraint solving seems a promising approach and it has been done with varying degrees of success. Early attempts to parallelize constraint propagation, which constitutes the core of traditional interleaved propagation and search constraint solving, were hindered by its essentially sequential nature. Recently, parallelization efforts have focussed mainly on the search part of constraint solving, as well as on local-search based solving. Lately, a particular source of parallelism has become pervasive, in the guise of GPUs, able to run thousands of parallel threads, and they have naturally drawn the attention of researchers in parallel constraint solving.

In this paper, we address challenges faced when using multiple devices for constraint solving, especially GPUs, such as deciding on the appropriate level of parallelism to employ, load balancing and inter-device communication, and present our current solutions.

Keywords: Constraint solving · Parallelism · GPU · Intel MIC
Hybrid systems

1 Introduction

Constraint Satisfaction Problems (CSPs) allow modeling problems like the Costas Array problem [6], and some real life problems like planning and scheduling [2], resources allocation [7] and route definition [3].

CPU's parallelism is already being used with success to speed up the solving processes of harder CSPs [5,16,19,21]. However, very few constraint solvers contemplate the use of GPUs. In fact, Jenkins *et al.* recently concluded that the execution model and the architecture of GPUs are not well suited to computations displaying irregular data access and code execution patterns such as backtracking search [10].

We are currently developing a constraint solver named Parallel Heterogeneous Architecture Toolkit (PHACT) that is already capable of achieving state-of-the-art performances on multi-core CPUs, and can also speed up the solving process by adding GPUs and processors like Intel Many Integrated Cores (MICs[1]) to solve the problems.

---

[1] Intel MICs are coprocessors that combine many Intel processor cores onto a single chip with dedicated RAM.

© Springer Nature Switzerland AG 2018
D. Seipel et al. (Eds.): DECLARE 2017, LNAI 10997, pp. 3–19, 2018.
https://doi.org/10.1007/978-3-030-00801-7_1

The next section introduces the main CSP concepts and Sect. 3 presents some related work. Section 4 describes the architecture of PHACT, and in Sect. 5 the results achieved with PHACT, when solving some CSPs on multiple combinations of devices and when compared with some state-of-the-art solvers, are displayed and discussed. Section 6 presents the conclusions and directions for future work.

2   CSPs Concepts

A CSP can be briefly described as a set of variables with finite domains, and a set of constraints between the values of those variables. The solution of a CSP is the assignment of one value from the respective domain to each one of the variables, ensuring that all constraints are met [3].

For example, the Costas Array problem consists in placing n dots on a $n \times n$ matrix such that each row and column contain only one dot and all vectors between dots are distinct. It can be modeled as a CSP with $n + n(n-1)/2$ variables, n of which correspond to the dots and each one is mapped to a different matrix column. The domain of these n variables is composed by the integers that correspond to the matrix rows where each dot may be placed. The remaining $n(n-1)/2$ variables constitute a difference triangle, whose rows cannot contain repeated values [6].

The methods for solving CSPs can be categorized as incomplete or complete. Incomplete solvers do not guarantee that an existing solution will be found, being mostly used for optimization problems and for large problems that would take too much time to fully explore. Incomplete search is beyond the scope of this paper and will not be discussed here. On the contrary, complete methods, such as the one implemented in PHACT, guarantee that if a solution exists, it will be found.

3   Related Work

Searching for CSP solutions in a backtracking approach can be represented in the form of a search tree. To take advantage of parallelism this search tree may be split into multiple subtrees and each one of them explored in a different thread that may be running on a different core, device or machine. This is the approach generally found in parallel constraint solvers, which run on single or distributed multi-core CPUs [5,16,19,21].

Pedro developed a CSP solver named Parallel Complete Constraint Solver (PaCCS) capable of running from a single core CPU to multiple multi-core CPUs in a distributed system [16]. Distributing the work among the threads through work stealing techniques and using the Message Passing Interface (MPI) to allow communication between machines, this solver achieved almost linear speedups for most of the problems tested, when using machines with up to 16 CPU cores.

Régin *et al.* implemented Embarrassingly Parallel Search, featuring an interface responsible for decomposing an initial problem into multiple sub-problems, filtering out those found to be inconsistent [20]. After generating the sub-problems it creates multiple threads, each one corresponding to an execution of a solver (e.g., Gecode [22]), to which a sub-problem is sent at a time for exploration.

For some optimization and search problems, where the full search space is explored, these authors achieved average gains of 13.8 and 7.7 against a sequential version, when using Gecode through their interface or just Gecode, respectively [20]. On their trials, the best results were achieved when decomposing the initial problem into 30 sub-problems per thread and running 40 threads on a machine with 40 CPU cores.

While solving CSPs through parallelization has been a subject of research for decades, the usage of GPUs for that purpose is a recent area, and as such there are not many published reports of related work. To our knowledge, there are only two published papers related with constraint solving on GPUs [1,4]. From these two, only Campeotto *et al.* presented a complete solver [4].

Campeotto *et al.* developed a CSP solver with Nvidia's Compute Unified Device Architecture (CUDA), capable of using simultaneously a CPU and an Nvidia GPU to solve CSPs [4]. On the GPU, this solver implements an approach different from the one mentioned before, namely, instead of splitting the search tree over multiple threads, it splits each constraint propagation over multiple threads. Constraints relating many variables are propagated on the GPU, while the remaining constraints are filtered sequentially by the CPU. On the GPU, the propagation and consistency check for each constraint are assigned to one or more blocks of threads according to the number of variables involved. The domain of each variable is filtered by a different thread.

Campeotto *et al.* reduced the data transfer to a minimum by transferring to the GPU only the domains of the variables that were not labeled yet and the events generated during the last propagation. Events identify the changes that happened to a domain, like becoming a singleton or having a new maximum value, which allows deciding on the appropriate propagator to apply.

Campeotto *et al.* obtained speedups of up to 6.61, with problems like the Langford problem and some real problems such as the modified Renault problem [4], when comparing a sequential execution on a CPU with the hybrid CPU/GPU version.

4 Solver Architecture

PHACT is a complete solver, capable of finding a solution for a CSP if one exists. It is meant to be able to use all the (parallel) processing power of the devices available on a system, such as CPUs, GPUs and MICs, to speed up solving constraint problems.

The solver is composed of a master process which collects information about the devices that are available on the machine, such as the number of cores and the type of device (CPU, GPU or MIC), and calculates the number of sub-search spaces that will be created to distribute among those devices. For each device there will be one thread (communicator) responsible for communicating with that device, and inside each device there will be a range of threads (search engines) that will perform labeling, constraint propagation and backtracking on one sub-search space at a time. The number of search engines that will be created inside each device will depend on the number of cores and type of that device, and may vary from 8 on a Quad-core CPU to more than 100,000 on a GPU.

PHACT may be used to count all the solutions of a given CSP, to find just one solution or a best one (for optimization problems).

Framework
PHACT is implemented in C and OpenCL [13], which allows its execution on multiple types of devices from different vendors and the capability of being executed on Linux or on Microsoft Windows.

We present some OpenCL concepts, in order to better understand PHACT's architecture:

- **Compute unit.** One or more processing elements and their local memory. In Nvidia GPUs each Streaming Multiprocessor (SM) is a compute unit. AMD GPUs have their own components called Compute Units that match this definition. For CPUs and MICs, the number of available compute units is normally equal to or higher than the number of threads that the device can execute simultaneously [13];
- **Kernel.** The code that will be executed on the devices;
- **Work-item.** An instance of the kernel (thread);
- **Work-group.** Composed of one or more work-items that will be executed on the same compute unit, in parallel. All work-groups for one kernel on one device have the same number of work-items;
- **Host.** CPU where the application responsible for managing the execution of the kernels is run;
- **Device.** A device where the kernels are executed (CPU, GPU, MIC).

In the implementation described here, the master process and the threads responsible for communicating with the devices run on the OpenCL host and the search engines run on the devices. The OpenCL host may also constitute a device, in which case it will be simultaneously controlling and communicating with the devices and running search engines. Each search engine corresponds to a work-item, and all work-items execute the same kernel code, which implements the search engine.

Search Space Splitting and Work Distribution
For distributing the work between the devices, PHACT splits the search space into multiple sub-search spaces. Search-space splitting is effected by partitioning the domains of one or more of the variables of the problem, so that the resulting sub-search spaces partition the full search space. The number and the size of the sub-search spaces thus created depend on the number of work-items which will be used, and may go up to a few millions.

Example 1 shows the result of splitting the search space of a CSP with three variables, $V1$, $V2$ and $V3$, all with domain $\{1,2\}$, into 4 sub-search spaces, $SS1$, $SS2$, $SS3$ and $SS4$.

Example 1. Creation of 4 sub-search spaces

$$SS1 = \{V1 = \{1\}, V2 = \{1\}, V3 = \{1,2\}\}$$
$$SS2 = \{V1 = \{1\}, V2 = \{2\}, V3 = \{1,2\}\}$$
$$SS3 = \{V1 = \{2\}, V2 = \{1\}, V3 = \{1,2\}\}$$
$$SS4 = \{V1 = \{2\}, V2 = \{2\}, V3 = \{1,2\}\}$$

Since each device will have multiple search engines running in parallel, the computed partition is organized into blocks of contiguous sub-search spaces that will be handled by each device, one at a time. The number of sub-search spaces that will compose each block will vary along the solving process and depends on the performance of each device on solving the current problem.

The communicator threads running on the host launch the execution of the search engines on the devices, hand each device one block of sub-search spaces to explore, and coordinate the progress of the solving process as each device finishes exploring its assigned block. The coordination of the devices consists in assessing the state of the search, distributing more blocks to the devices, signaling to all the devices that they should stop (when a solution has been found and only one is wanted), or updating the current bound (in optimization problems).

Load Balancing

An essential aspect to consider when parallelizing some task is the balancing of the work between the parallel components. Creating sub-search spaces with balanced domains, when possible, is no guarantee that the amount of work involved in exploring each of them is even similar. To compound the issue, we are dealing with devices with differing characteristics and varying speeds, making it even harder to statically determine an optimal, or even good, work distribution.

Achieving effective load balancing between devices with such different architectures as CPUs and GPUs is a complex task [10]. When trying to implement dynamic load balancing, two important OpenCL limitations arise, namely when a device is executing a kernel it is not possible for it to communicate with other devices [8], and the execution of a kernel can not be paused or stopped. Hence, techniques like work stealing [5,17], which requires communication between threads, will not work with kernels that run independently on different devices and load balancing must be done on the host side.

To better manage the distribution of work, the host could reduce the amount of work it sends to the devices each time, by reducing the number of sub-search spaces in each block. This would make the devices synchronize more frequently on the host and allow for a finer control over the behavior of the solver. When working with GPUs, though, the number and the size of data transfers between the devices and the host should be as small as possible, because these are very time consuming operations. So, a balance must be struck between the workload of the devices and the amount of communication needed.

PHACT implements a dynamic load balancing technique which adjusts the size of the blocks of sub-search spaces to the performance of each device solving the current problem, when compared to the performance of the other devices.

Initially each device d explores two small blocks of sub-search spaces to get the *average time*, $avg(d)$, it needs to explore one sub-search space. The size of those blocks may be distinct among devices as it is calculated according to the number of threads that each device is capable of running simultaneously and its clock speed. When two or more devices finish exploring those first two blocks, their *rank*, $rank(d)$ is calculated according to Eq. (1), where m is the total number of devices.

$$rank(d) = \frac{\frac{1}{avg(d)}}{\sum_{i=1}^{m} \frac{1}{avg(i)}}, \quad avg(i) > 0 \tag{1}$$

The rank of a device consists of a value between 0 and 1, corresponding to the relative speed of the device against all the devices that were used for solving a block of sub-search spaces. Faster devices will get a higher rank than slower devices, and the sum of the ranks of all the devices will be 1. The rank is then used to calculate the size of the next block of sub-search spaces to send to the device, by multiplying its value by the number of sub-search spaces that are yet to be explored.

Since the size of the first two blocks of sub-search spaces explored by each device is small, to prevent slow devices from dominating the solving process, it often only allows for a rough approximation of the speed of a device. So, in the beginning, only 1/3 of the remaining sub-search spaces are considered when computing the size of the next block to send to a device.

For the first device to finish its first two blocks, it will not be possible to calculate its rank, as it would need the *average time* of at least one more device. In this case, that device will get a new block with twice the size of the previous ones, as this device is probably the fastest device solving the current problem.

As search progresses, every time a device finishes exploring another block, its average time and rank are updated. The value of the average time of a device is the result of dividing the total time that the device was exploring sub-search spaces by the total number of sub-search spaces that it explored already.

As the rank value stabilizes, the size of the new block of sub-search spaces for the device will be the corresponding percentage from all unexplored sub-search spaces. Table 1 exemplifies the calculation of the number of search spaces that will compose the block of search spaces which will be sent for each device as soon as each of them finishes its previous block. This is repeated until a device waiting for work is estimated to need less than one second[2] to solve all the remaining sub-search spaces, in which case it will be assigned all of them.

---

[2] If a device takes less than one second to explore a block of search spaces, most of that time is spent communicating with the host and initializing its data structures.

Table 1. Example of the calculation of blocks size when using three devices

Device	Average time per search space (ms)	Rank	Remaining sub-search spaces to explore	Size of the next block of sub-search spaces
Device 1	0.00125	0.625	1233482	770926
Device 2	0.00236	0.331	462556	153106
Device 3	0.01782	0.044	309450	13616

Another challenge GPUs pose is that they achieve the best performance when running hundreds or even thousands of threads simultaneously. But to use that level of parallelism, they must have enough work to keep that many threads busy. Otherwise, when a GPU receives a block with less sub-search spaces than the number of threads that would allow it to achieve its best performance, the average time needed to explore one sub-search space increases sharply.

For example the Nvidia GeForce GTX 980M takes about 1.1 s to find all the solutions for the n-Queens 13 when splitting the problem in 742,586 sub-search spaces, and 2.4 s when split in only 338 sub-search spaces. This challenge is also valid for CPUs, but not so problematic due to their lesser degree of parallelism when compared with the GPUs.

To overcome that challenge, sub-search spaces may be further divided inside a device, by applying a multiplier factor m to the size of a block and turning a block of sub-search spaces into a block with m times the original number of sub-search spaces, that will be created as presented in Example 1.

Communication

To reduce the amount of data that is transferred to each device, all of them will receive the full CSP, that is, all the constraints, variables and their domains, at the beginning of the solving process. Afterwards, when a device must be instructed to solve a new block of sub-search spaces, instead of sending all the sub-search spaces to the device, only the information needed to create those sub-search spaces is sent.

If a device is to solve sub-search spaces $SS2$ and $SS3$ from Example 1, it will receive the information that the tree must be expanded down to depth 2, that the values of the first variable are repeated 2 times and the values of the second variable are repeated 1 time only (not repeated). With this information the device will know that the values of the first variable are repeated 2 times, so the third sub-search space ($SS3$) will get the second value of that variable, and so on down to the expansion depth. The values of the variables that were not expanded are simply copied from the original CSP that was passed to the devices at the beginning of the solving process.

Each time a work-item needs a new sub-search space to explore, it increases by one the number of the first/next sub-search space that is yet to be explored on that device and creates the sub-search space corresponding to the number before being increased. Then it will do labeling, propagation and backtracking on that search-space, repeating all these steps until either all the sub-search spaces of that block have been explored, when all the solutions must be found, or only one solution is wanted and one of the work-items on that device finds a solution.

Implementation Details

Several tests were made to find the best number of work-groups to use for each type of device. It was found that for CPUs and MICs the best results were achieved with the same number of work-groups as the amount of compute units of the device. For GPUs, the predefined number of work-groups is 4096 due to the increased level of parallelism allowed by this type of devices.

The user can specify how many sub-search spaces must be created or let PHACT estimate that number. For estimating the number of sub-search spaces that will be generated, PHACT will sum all the work-items that will be used in all the devices and multiply that value by 40 if all the solutions must be found for the current CSP, or by 100 if only one solution is required or when solving an optimization problem. After several tests these values (40 and 100) were found as allowing to achieve a good load balancing between the devices, and as such they are the predefined values.

When looking for just one solution or optimizing, the amount of work sent to each device is reduced by generating more sub-search spaces and decreasing the size of the blocks sent to the devices, which makes each one of them faster to explore, to make sure all the devices are synchronized on the host more frequently.

As for the number of work-items per work-group, CPUs and MICs are assigned one work-item per work-group, as their compute units can only execute one thread at a time.

On the contrary, each GPU compute unit can execute more than one thread simultaneously. For example, the Nvidia GeForce GTX 980 has 16 SMs with 128 CUDA cores[3] each, making a total of 2048 CUDA cores. Nevertheless, each SM is only capable of executing simultaneously 32 threads (using only 32 CUDA cores at the same time) making it capable of running 512 threads simultaneously [15].

Each SM has very limited resources that are shared between work-groups and their work-items, thus limiting the number of work-items per work-group that can be used according to the resources needed by each work-item. The main limitation is the size of the local memory of each SM that is shared between all the work-items of the same work-group and between some work-groups (8 work-groups for the Nvidia GeForce GTX 980).

For this reason, PHACT estimates the best number of work-items per work-group to use for GPUs, by limiting the amount of local memory required to the size of the available local memory on the GPU. When the available local memory is not enough to efficiently use at least one work-item per work-group, PHACT will only use the global memory of the device, which is much larger but also much slower, and 32 work-items per work-group, as each SM is only capable of running 32 threads simultaneously.

---

[3] A CUDA core is a processing element capable of executing one integer or floating instruction per clock for a thread.

Note that PHACT represents variable domains as either 32-bit bitmaps, multiples of 64-bit bitmaps, or as (compact) intervals. When using intervals, PHACT is slower than when using bitmaps, but intervals are meant to be used instead of larger bitmaps on systems where the size of the RAM is an issue.

The techniques described in this section allow PHACT to use all the devices compatible with OpenCL to solve a CSP. It splits the search space in multiple search spaces that are distributed among the devices in blocks to reduce the number of communications between the host and the devices. The size of each block is calculated according to the speed of the respective device when solving the previous blocks to try to achieve a good load balancing between the devices. The size of the data transfers between the devices and the host is reduced by replacing the blocks of fully created search spaces with a small data set containing the information needed for a device to generate those search spaces.

5   Results and Discussion

PHACT was evaluated on finding all the solutions for four different CSPs, on optimizing one other CSP and on finding one solution for another CSP, each one with two different sizes, except for the Latin Problem whose smaller size is solved too fast and a bigger size takes too long to solve. Those tests were executed on one, two and three devices and on four different machines running Linux to evaluate the speedups when adding more devices to help the CPU.

PHACT performance was compared with those of PaCCS and Gecode 5.1.0 on these four machines. The four machines have the following characteristics:

M1. Machine with 32 GB of RAM and:
  - Intel Core i7-4870HQ (8 compute units);
  - Nvidia GeForce GTX 980M (12 compute units).

M2. Machine with 64 GB of RAM and:
  - Intel Xeon E5-2690 v2 (referred to as Xeon 1, 40 compute units);
  - Nvidia Tesla K20c (13 compute units).

M3. Machine with 128 GB of RAM and:
  - AMD Opteron 6376 (64 compute units);
  - Two AMD Tahitis (32 compute units each). These two devices are combined in an AMD Radeon HD 7990, but are managed separately by OpenCL.

M4. Machine with 64 GB of RAM and:
  - Intel Xeon CPU E5-2640 v2 (referred to as Xeon 2, 32 compute units);
  - Two Intel Many Integrated Core 7120P (240 compute units each).

Tables 2, 3, 4 and 5 present the elapsed times and speedups when solving all the problems on M1, M2, M3 and M4, respectively. Five of the six problems models were retrieved from the Minizinc Benchmarks suite [12]. The Langford Numbers problem was retrieved from CSPLib [9], due to the absence of reified constraints on PHACT and PaCCS, that are used in the Minizinc Benchmarks model, which would lead to different constraints being used among the three solvers. PaCCS does not have the "absolute value" constraint implemented, so it was not tested with the All Interval problem.

This set of problems allowed to evaluate the solvers with 8 different constraints combined with each other in different ways. All the solutions were found for the problems whose name is followed by "(Count)" on the tables, the optimal solution was searched for the problem identified with "(Optim.)" and for the problem whose name is followed by "(One)", only one solution was required.

For simplicity, the 4 tables have the resources used on the respective machine identified as R1, R2, R3 and R4, where R1 means using only a single thread on the CPU, R2 means using all the threads of that CPU, R3 means using all the threads on the CPU and one device (Geforce, Tesla, Tahiti or MIC), and R4 means using all the threads on the CPU and two identical devices (MICs or Tahitis). It must be noted that only PHACT is capable of using R3 and R4 resources.

Table 2 shows that using the Geforce to help I7 allowed speedups of up to 4.66. However, in two problems, using also the Geforce resulted in more time needed to solve the same problems. This result is mainly due to the small number of work-items per work-group that was effectively used on Geforce, due to the local memory limitations detailed in Sect. 4. On this machine, adding the Geforce to help I7 allowed a geometric mean speedup of 1.53.

The slowdown noted when optimizing the Golomb Ruler with 12 marks is also due to the impossibility of different devices communicating with each other while their kernels are running, as stated in Sect. 4. This is problematic when optimizing, as a device which finds a better solution cannot tell the other devices to find only solutions better than the one it just found. Instead it will finish exploring its block of sub-search spaces and only after that it will inform the host about the new solution, and only after this point, when another device finishes its block, it will be informed about the new solution that must be optimized. Due to this limitation, the devices spend some time looking for solutions that may already be worse than the ones found by other devices. This problem was also noted on the other three machines.

As for the Langford Numbers problem with 14 numbers, the worse result when adding the Geforce was due to the very unbalanced sub-search spaces that are generated leading to most of sub-search spaces being easily detected as inconsistent, and only a few containing most of the work. This is problematic, because as each thread explores each sub-search space sequentially, in the end only a few threads will be working on the harder sub-search spaces while the others are idle. This problem was also noted on the other three machines.

PHACT was faster than PaCCS in all problems, achieving speedups of up to 5.37.

When comparing with Gecode, PHACT achieved good speedups on all the problems, except on Market Split, which is a simple problem with only one constraint type which may have a faster propagator on Gecode. On the contrary, with the Latin problem, Gecode was 127.85 times slower than PHACT when

Table 2. Elapsed times and speedups on M1, with 4 cores and 1 GPU

CSP	Resources	PHACT		PaCCS			GECODE		
		Elapsed (s)	Speedup vs. fewest resources	Elapsed (s)	Speedup vs. fewest resources	PHACT speedup	Elapsed (s)	Speedup vs. fewest resources	PHACT speedup
All Interval 14	R1	872.31					1188.10		1.36
(Count)	R2	101.17	8.62				304.78	3.90	3.01
All Interval 15	R2	1477.15					3303.50		2.24
(Count)	R3	317.32	4.66						10.41
Costas Array 13	R1	70.64		149.68		2.12	180.30		2.55
(Count)	R2	16.41	4.30	34.85	4.29	2.12	41.78	4.32	2.55
Costas Array 15	R2	555.61		1295.94		2.33	1422.73		2.56
(Count)	R3	409.24	1.36			3.17			3.48
Golomb Ruler	R1	172.08		498.32		2.90	366.65		2.13
11 (Optim.)	R2	41.08	4.19	122.96	4.05	2.99	91.54	4.01	2.23
Golomb Ruler	R2	454.43		1440.14		3.17	1148.96		2.53
12 (Optim.)	R3	468.55	0.97			3.07			2.45
Langford Numb.	R1	64.85		76.99		1.19	104.96		1.62
13 (Count)	R2	19.54	3.32	21.04	3.66	1.08	24.84	4.23	1.27
Langford Numb.	R2	159.58		184.11		1.15	197.97		1.24
14 (Count)	R3	171.65	0.93			1.07			1.15
Latin 6 (Count)	R1	1010.30		1746.46		1.73	20936.00		20.72
	R2	209.40	4.82	385.62	4.53	1.84	26771.00	0.78	127.85
	R3	118.13	1.77			3.26			226.62
Market Split	R1	10.37		11.04		1.06	9.69		0.93
s4-07 (One)	R2	3.16	3.28	1.19	9.28	0.38	3.94	2.46	1.25
Market Split	R2	212.86		897.21		4.22	159.94		0.75
s5-01 (One)	R3	167.07	1.27			5.37			0.96

using only the CPU. Gecode was slower in solving this problem with all the CPU threads than when using only one thread, which suggests that the method used for load balancing between threads is very inefficient for this problem. This behavior of Gecode was noted in all the machines.

Table 3 presents the results on solving the same problems on M2. Using the Tesla GPU to help the Xeon 1 resulted in most of the cases in a slowdown. In fact, adding the Tesla to help Xeon 1 introduced an average slowdown of 0.84. This is due to the fact that Tesla was the slowest GPU used on the tests, being no match for Xeon 1. In fact, the work done by Tesla did not compensate the time spent by Xeon 1 (host) to control Tesla (device).

On this machine, PHACT was faster than PaCCS in all but one problem, resulting in an average speedup of 1.44 favoring PHACT. Comparing with Gecode, PHACT was faster on all the problems with all the resources combinations.

The results for the M3 machine are presented in Table 4. This machine possesses the CPU used on the tests that has the greater number of cores (64), and it is paired up with two Tahiti GPUs, that are faster than Tesla, but slower than Geforce. So it is very hard for the Tahitis to display some performance gains when compared with a 64 cores CPU. However, with the All Interval 15 problem, they were capable of speeding up the solving process by 1.48 times. On average, adding the two Tahiti GPUs to help Opteron did not allow any

Table 3. Elapsed times and speedups on M2, with 40 cores and 1 GPU

CSP	Resources	PHACT		PaCCS			GECODE		
		Elapsed (s)	Speedup vs. fewest resources	Elapsed (s)	Speedup vs. fewest resources	PHACT speedup	Elapsed (s)	Speedup vs. fewest resources	PHACT speedup
All Interval 14	R1	920.77					1356.75		1.47
(Count)	R2	24.25	37.97				236.19	5.74	9.74
All Interval 15	R2	329.57					2294.76		6.96
(Count)	R3	263.60	1.25						8.71
Costas Array 13	R1	74.77		183.80		2.46	188.43		2.52
(Count)	R2	4.79	15.61	7.83	23.47	1.63	10.35	18.21	2.16
Costas Array 15	R2	108.17		279.78		2.59	327.80		3.03
(Count)	R3	144.73	0.75			1.93			2.26
Golomb Ruler	R1	187.59		548.46		2.92	363.87		1.94
11 (Optim.)	R2	10.59	17.71	27.84	19.70	2.63	25.85	14.08	2.44
Golomb Ruler	R2	114.22		253.73		2.22	293.17		2.57
12 (Optim.)	R3	180.56	0.63			1.41			1.62
Langford Numb.	R1	78.83		83.84		1.06	115.89		1.47
13 (Count)	R2	6.52	12.09	4.25	19.73	0.65	13.52	8.57	2.07
Langford Numb.	R2	41.89		36.41		0.87	100.98		2.41
14 (Count)	R3	60.24	0.70			0.60			1.68
Latin 6 (Count)	R1	1126.44		1849.59		1.64	26653.00		23.66
	R2	51.24	21.98	85.25	21.70	1.66	59227.00	0.45	1155.87
	R3	56.77	0.90			1.50			1043.28
Market Split	R1	11.17		12.24		1.10	11.22		1.00
s4-07 (One)	R2	2.37	4.71	0.30	40.80	0.13	5.94	1.89	2.51
Market Split	R2	46.39		148.69		3.21	1163.47		25.08
s5-01 (One)	R3	48.09	0.96			3.09			24.19

speedup, because the time spent by Opteron to control and communicate with the Tahitis was similar to the time that the Opteron would take to perform the work done by the Tahitis.

The issues with Golomb Ruler and Langford Number discussed before in this section, were also noted on this machine.

When comparing with PaCCS, PHACT achieved speedups that ranged from 0.21 on a very small problem to 4.67. PHACT was faster than Gecode in all the tests, except when optimizing Golomb Ruler 12 with the Opteron and one Tahiti.

Table 5 presents the results on the M4 machine. This machine possesses two MICs whose architecture is more similar to the CPUs than to GPUs, so, they are more prepared for solving sequential problems than GPUs. That difference was noted with the Langford Numbers problem, where they were capable of achieving a speedup of 1.51 despite the unbalanced sub-search spaces. On this machine, adding the two MICs to help Xeon 2 allowed an average speedup of 1.45. When counting all the solutions for the Costas Array 15, the two MICs allowed a top speedup of 1.90.

When compared with PaCCS and Gecode the results are very similar to the ones achieved on the other machines, being faster than Gecode in all but one problem and faster than PaCCS in 19 of the 24 tests.

Table 4. Elapsed times and speedups on M3, with 64 cores and 2 GPUs

CSP	Resources	PHACT		PaCCS			GECODE		
		Elapsed (s)	Speedup vs. fewest resources	Elapsed (s)	Speedup vs. fewest resources	PHACT speedup	Elapsed (s)	Speedup vs. fewest resources	PHACT speedup
All Interval 14 (Count)	R1	1806.78					2363.88		1.31
	R2	26.46	68.28				838.62	2.82	31.69
All Interval 15 (Count)	R2	351.33					8502.00		24.20
	R3	175.26	2.00						48.51
	R4	237.49	1.48						35.80
Costas Array 13 (Count)	R1	133.62		345.57		2.59	334.95		2.51
	R2	5.09	26.25	7.91	43.69	1.55	31.39	10.67	6.17
Costas Array 15 (Count)	R2	114.31		281.16		2.46	692.10		6.05
	R3	118.13	0.97			2.38			5.86
	R4	111.69	1.02			2.52			6.20
Golomb Ruler 11 (Optim.)	R1	378.81		1225.71		3.24	640.88		1.69
	R2	11.56	32.77	29.72	41.24	2.57	26.80	23.91	2.32
Golomb Ruler 12 (Optim.)	R2	126.72		309.72		2.44	312.81		2.47
	R3	361.60	0.35			0.86			0.87
	R4	138.28	0.92			2.24			2.26
Langford Numb. 13 (Count)	R1	133.78		173.59		1.30	195.70		1.46
	R2	6.38	20.97	4.85	35.79	0.76	47.95	4.08	7.52
Langford Numb. 14 (Count)	R2	40.42		40.91		1.01	387.10		9.58
	R3	101.43	0.40			0.40			3.82
	R4	56.11	0.72			0.73			6.90
Latin 6 (Count)	R1	1871.64		3489.64		1.86	49361.00		26.37
	R2	50.07	37.38	79.12	44.11	1.58	138065.00	0.36	2757.44
	R3	56.95	0.88			1.39			2424.32
	R4	44.37	1.13			1.78			3111.67
Market Split s4-07 (One)	R1	19.24		23.24		1.21	20.54		1.07
	R2	2.25	8.55	0.48	48.42	0.21	21.81	0.94	9.69
Market Split s5-01 (One)	R2	46.69		218.07		4.67	2505.31		53.66
	R3	48.48	0.96			4.50			51.68
	R4	50.96	0.92			4.28			49.16

Figure 1 presents the geometric mean of the speedups achieved by PHACT against PaCCS and Gecode, showing that PHACT was faster than Gecode and PaCCS on all the machines with all the resources combinations.

We can observe that the difference in performance between PHACT and Gecode is greater on the machines that have a CPU with more cores, which shows that the load balancing techniques implemented in PHACT are more efficient for the problems that were presented here. When compared with PaCCS, that relation is no longer noticed and the results are much closer between the two solvers when using only the CPUs.

Using all the available resources on the four machines allowed PHACT to increase its performance when compared to PaCCS and Gecode, which shows that its greater versatility can lead to an improved performance.

Table 5. Elapsed times and speedups on M4, with 32 cores and 2 MICs

CSP	Resources	PHACT		PaCCS			GECODE		
		Elapsed (s)	Speedup vs. fewest resources	Elapsed (s)	Speedup vs. fewest resources	PHACT speedup	Elapsed (s)	Speedup vs. fewest resources	PHACT speedup
All Interval 14 (Count)	R1	1697.78					1940.32		1.14
	R2	43.69	38.86				310.89	6.24	7.12
All Interval 15 (Count)	R2	624.58					2986.63		4.78
	R3	304.83	2.05						9.80
	R4	245.43	2.54						12.17
Costas Array 13 (Count)	R1	118.40		262.07		2.21	270.49		2.28
	R2	7.73	15.32	13.92	18.83	1.80	17.62	15.35	2.28
Costas Array 15 (Count)	R2	229.28		507.47		2.21	586.08		2.56
	R3	174.52	1.31			2.91			3.36
	R4	120.85	1.90			4.20			4.85
Golomb Ruler 11 (Optim.)	R1	262.06		793.89		3.03	526.13		2.01
	R2	17.35	15.10	48.15	16.49	2.78	42.85	12.28	2.47
Golomb Ruler 12 (Optim.)	R2	195.73		439.97		2.25	456.06		2.33
	R3	388.82	0.50			1.13			1.17
	R4	279.53	0.70			1.57			1.63
Langford Numb. 13 (Count)	R1	119.68		120.81		1.01	166.89		1.39
	R2	10.48	11.42	7.51	16.09	0.72	17.33	9.63	1.65
Langford Numb. 14 (Count)	R2	79.72		64.10		0.80	132.85		1.67
	R3	77.60	1.03			0.83			1.71
	R4	52.76	1.51			1.21			2.52
Latin 6 (Count)	R1	1645.52		2664.03		1.62	39611.00		24.07
	R2	84.38	19.50	149.60	17.81	1.77	72203.00	0.55	855.69
	R3	73.69	1.15			2.03			979.82
	R4	65.59	1.29			2.28			1100.82
Market Split s4-07 (One)	R1	18.35		17.54		0.96	15.94		0.87
	R2	2.48	7.40	0.43	40.79	0.17	7.34	2.17	2.96
Market Split s5-01 (One)	R2	80.88		213.38		2.64	280.07		3.46
	R3	68.03	1.19			3.14			4.12
	R4	57.96	1.40			3.68			4.83

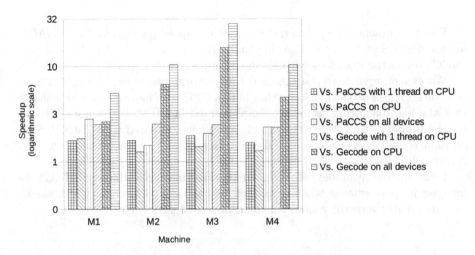

Fig. 1. Speedups when using PHACT against PaCCS and Gecode on the four machines

6   Conclusion and Future Work

To our knowledge, PHACT is the only constraint solver capable of using simultaneously CPUs, GPUs, MICS and any other device compatible with OpenCL to solve CSPs in a faster manner. Although GPUs are not particularly efficient for this type of problems, they still can speed up the solving process and in some cases, be even faster than the CPU of the same machine.

PHACT has been tested with 6 different CSPs on 4 different machines with 2 and 3 devices each, namely Intel CPUs and MICs, Nvidia GPUs, and AMD CPUs and GPUs, allowing it to achieve speedups of up to 4.66 when compared with using only the CPU of the machine to solve a single CPS, and a geometric mean speedup of up to 1.53 when solving all the referred CSPs on each machine.

On the four machines used for testing, PHACT achieved a geometric mean speedup that ranged from 1.28 to 2.83 when compared with PACCS, and 2.31 to 28.44 when compared with Gecode. The use of all the devices compatible with OpenCL to solve a CSP allowed PHACT to improve its performance against PaCCS and Gecode when compared with using only the CPUs.

Campeotto *et al.* [4] achieved a top speedup of 6.61 when using one thread of a CPU together with a GPU, while PHACT achieved average speedups between 1.56 and 2.88 when using one thread on a CPU and a GPU, with a top speedup of 15.15, and average and top speedups of 7.33 and 30.63 when replacing the GPU by two MICs. Although their technique of using the GPUs to propagate constraints relating many variables seems to have significant host–device synchronization requirements, we intend to test this approach in the future.

PHACT is yet being improved to try to overcome the lack of synchronization between devices when optimizing. The solution may pass by more frequent communication between host and devices, taking into account the number of solutions already found and increasing the frequency of the communication for problems with more solutions.

As for the unbalanced sub-search spaces that lead to only a few threads working in parallel while the others have already finished their work, we are analysing a work-sharing strategy [18] that may be executed when all the sub-search spaces generated for the block have ended but some threads are still working.

A MiniZinc/FlatZinc [14] reader is also being implemented to allow the direct input of problems already modeled in this language.

Acknowledgments. This work was partially funded by Fundação para a Ciência e Tecnologia (FCT) under grant UID/CEC/4668/2016 (LISP). Some of the experimentation was carried out on the `khromeleque` cluster of the University of Évora, which was partly funded by grants ALENT-07-0262-FEDER-001872 and ALENT-07-0262-FEDER-001876.

References

1. Arbelaez, A., Codognet, P.: A GPU implementation of parallel constraint-based local search. In: 2014 22nd Euromicro International Conference on PDP, pp. 648–655. IEEE (2014)
2. Barták, R., Salido, M.A.: Constraint satisfaction for planning and scheduling problems. Constraints 16(3), 223–227 (2011)
3. Brailsford, S., Potts, C., Smith, B.: Constraint satisfaction problems: algorithms and applications. Eur. J. Oper. Res. 119, 557–581 (1999)
4. Campeotto, F., Dal Palù, A., Dovier, A., Fioretto, F., Pontelli, E.: Exploring the use of GPUs in constraint solving. In: Flatt, M., Guo, H.-F. (eds.) PADL 2014. LNCS, vol. 8324, pp. 152–167. Springer, Cham (2014). https://doi.org/10.1007/978-3-319-04132-2_11
5. Chu, G., Schulte, C., Stuckey, P.J.: Confidence-based work stealing in parallel constraint programming. In: Gent, I.P. (ed.) CP 2009. LNCS, vol. 5732, pp. 226–241. Springer, Heidelberg (2009). https://doi.org/10.1007/978-3-642-04244-7_20
6. Diaz, D., Richoux, F., Codognet, P., Caniou, Y., Abreu, S.: Constraint-based local search for the costas array problem. In: Hamadi, Y., Schoenauer, M. (eds.) LION 2012. LNCS, pp. 378–383. Springer, Heidelberg (2012). https://doi.org/10.1007/978-3-642-34413-8_31
7. Filho, C., Rocha, D., Costa, M., Albuquerque, P.: Using constraint satisfaction problem approach to solve human resource allocation problems in cooperative health services. Expert Syst. Appl. 39(1), 385–394 (2012)
8. Gaster, B., Howes, L., Kaeli, D., Mistry, P., Schaa, D.: Heterogeneous Computing with OpenCL. Morgan Kaufmann Publishers Inc., San Francisco (2011)
9. Jefferson, C., Miguel, I., Hnich, B., Walsh, T., Gent, I.P.: CSPLib: a problem library for constraints (1999). http://www.csplib.org
10. Jenkins, J., Arkatkar, I., Owens, J.D., Choudhary, A., Samatova, N.F.: Lessons learned from exploring the backtracking paradigm on the GPU. In: Jeannot, E., Namyst, R., Roman, J. (eds.) Euro-Par 2011. LNCS, vol. 6853, pp. 425–437. Springer, Heidelberg (2011). https://doi.org/10.1007/978-3-642-23397-5_42
11. Mairy, J.-B., Deville, Y., Lecoutre, C.: Domain k-wise consistency made as simple as generalized arc consistency. In: Simonis, H. (ed.) CPAIOR 2014. LNCS, vol. 8451, pp. 235–250. Springer, Cham (2014). https://doi.org/10.1007/978-3-319-07046-9_17
12. MIT: a suite of minizinc benchmarks (2017). https://github.com/MiniZinc/minizinc-benchmarks
13. Munshi, A., Gaster, B., Mattson, T.G., Fung, J., Ginsburg, D.: OpenCL Programming Guide, 1st edn. Addison-Wesley Professional, Boston (2011)
14. Nethercote, N., Stuckey, P.J., Becket, R., Brand, S., Duck, G.J., Tack, G.: MiniZinc: towards a standard CP modelling language. In: Bessière, C. (ed.) CP 2007. LNCS, vol. 4741, pp. 529–543. Springer, Heidelberg (2007). https://doi.org/10.1007/978-3-540-74970-7_38
15. NVIDIA Corporation: NVIDIA GeForce GTX 980 featuring maxwell, the most advanced GPU ever made. White paper. NVIDIA Corporation (2014)
16. Pedro, V.: Constraint programming on hierarchical multiprocessor systems. Ph.D. thesis, Universidade de Évora (2012)
17. Pedro, V., Abreu, S.: Distributed work stealing for constraint solving. In: Vidal, G., Zhou, N.F. (eds.) CICLOPS-WLPE 2010, Edinburgh, Scotland, U.K. (2010)

18. Rolf, C.C., Kuchcinski, K.: Load-balancing methods for parallel and distributed constraint solving. In: 2008 IEEE International Conference on Cluster Computing, pp. 304–309, September 2008
19. Rolf, C.C., Kuchcinski, K.: Parallel solving in constraint programming. In: MCC 2010, November 2010
20. Régin, J.-C., Rezgui, M., Malapert, A.: Embarrassingly parallel search. In: Schulte, C. (ed.) CP 2013. LNCS, vol. 8124, pp. 596–610. Springer, Heidelberg (2013). https://doi.org/10.1007/978-3-642-40627-0_45
21. Schulte, C.: Parallel search made simple. In: Beldiceanu, N., et al. (eds.) Proceedings of TRICS: CP 2000, Singapore, September 2000
22. Schulte, C., Duchier, D., Konvicka, F., Szokoli, G., Tack, G.: Generic constraint development environment. http://www.gecode.org/

Run-Time Analysis of Temporal Constrained Objects

Jinesh M. Kannimoola[1]([✉]), Bharat Jayaraman[2], and Krishnashree Achuthan[1]

[1] Center for Cybersecurity Systems and Networks,
Amrita Vishwa Vidyapeetham, Amritapuri, Kollam, India
jinesh@am.amrita.edu, krishna@amrita.edu
[2] Department of Computer Science and Engineering,
State University of New York at Buffalo, Buffalo, USA
bharat@buffalo.edu

Abstract. The programming paradigm of constrained objects is a declarative variant of the object-oriented paradigm wherein objects define the structure of a system and declarative constraints (rather than imperative methods) define its behavior. Constrained objects have many uses in the engineering domain and computation in this paradigm is essentially constraint solving. This paper is concerned with an extension of constrained objects called temporal constrained objects, which are especially appropriate for modeling dynamical systems. The main extensions are series variables and metric temporal operators to declaratively specify time-varying behavior. The language TCOB exemplifies this paradigm and the execution of TCOB programs consists of constraint solving within a time-based simulation framework. One of the challenges in TCOB is identifying errors owing both to the complexity of programs and the underlying constraint solving methods. We address this problem by extracting a run-time trace of the execution of a TCOB program and providing an analysis of the cause of error. The run-time trace also serves as a basis, in many cases, for constructing a finite-state machine which in turn can be used for 'model-checking' properties of the system. The paper also presents abstraction techniques for dealing with simulations that result in large state spaces.

Keywords: Temporal constraints objects · Time-based simulation
Run-time verification · Finite state models · Error analysis
Predicate abstraction · Visualization

1 Introduction

Constrained objects are a natural modeling approach for complex structures with two essential characteristics: (i) They are compositional in nature, i.e., a complex structure is built (recursively) of smaller structures. (ii) The behavior of an individual component by itself and its relation to other components are

© Springer Nature Switzerland AG 2018
D. Seipel et al. (Eds.): DECLARE 2017, LNAI 10997, pp. 20–36, 2018.
https://doi.org/10.1007/978-3-030-00801-7_2

regulated by laws, or rules. The language COB exemplifies this concept, and it has been shown to be useful especially in modeling complex engineering structures [7]. COB makes use of Java-like classes, inheritance, and aggregation for modeling structure and makes use of declarative constraints (rather than imperative methods) for modeling behavior. The emergent behavior of a collection of constrained objects is determined by a process of constraint solving over the attributes of objects.

Temporal constrained objects [8] are an extension of constrained objects that are particularly suited for modeling the time-dependent behavior of complex dynamic systems. The new feature here is the *series variable*, which records the sequence of changes to some entity of interest in a dynamic system. Time is considered as a metric quantity; a built-in variable `Time` represents the current time and it is automatically incremented by one unit to record the passage of time. For example, while the current and voltage across a resistor in a DC circuit can be modeled using ordinary variables, in an AC circuit the current and voltage change with time and hence are better modeled with series variables. It is common, in such examples, for constraints to be placed over consecutive values in the time-sequence. The language TCOB extends COB with series variables as well as *metric temporal operators*, which are a metric variant of the classic temporal operators of LTL [3]. Together they are effective in specifying the overall dynamic behavior of a variety of complex structures. The execution model for TCOB involves a time-based simulation along with constraint solving at each time-step.

Run-time analysis refers to methods and tools for monitoring the run-time behavior of a program or system with the goal of debugging and verifying its behavior. An important aspect of run-time analysis is run-time verification, which attempts to bridge the gap between formal verification and software testing [9]. Most of these techniques are based upon a finite execution trace of a program [13]. The execution trace records the major events that occurred during execution, such as variable/field read's and write's, method call/return, and object creation. In this paper we investigate the usability of a similar approach in the paradigm of temporal constrained objects, an important difference being that state updating is not permitted in our paradigm.

The time-based simulation used in temporal constrained objects naturally lends itself to analysis based upon a linear execution trace. Detecting errors when constraint solving is interwoven with a time-based simulation is especially challenging because the cause of an error often is often separated by many time steps from the point at which the symptom of the error manifests. We show how run-time analysis together with run-time visualization greatly help in addressing this challenge. We construct a temporal constraint dependency graph at run-time in order to clarify temporal dependencies and help identification of errors.

We also extract a *finite state machine* from the execution trace and formulate properties of interest as verification conditions in a propositional temporal logic [3]. A state consists of the values of a set of variables chosen by the user. The set of states is the set of distinct combinations of values taken by these variables during the course of program execution. Since state updating is not possible in temporal constrained objects, state changes are possible only because series variables may assume different values as time progresses. Of course, it is possible that the series variables assume the same values at two different points in time, i.e., it is possible that states repeat. Sometimes there could be a large number of states, and there is a need to construct a reduced run-time model that clarifies the emergent behavior at a high level. We propose an approach which we call *predicate abstraction* in order to reduce the number of states without losing an abstract view of the system. In our approach, we can directly encode the predicate abstraction rules as constraints in the system. Thus, our run-time analysis is a combination of visualization, error-detection and verification.

The remainder of this paper is organized as follows. Section 2 discusses the related literature in this field; Sect. 3 introduces the concept of temporal constrained objects with the aid of examples; Sect. 4 presents the run-time analysis of temporal constrained objects; and, finally, Sect. 5 presents conclusions and areas of further work.

2   Related Work

Run-time verification is based on extracting a trace from a running system and using it to detect observed behaviors satisfying or violating certain properties [9]. Binary code instrumentation [4] is one of the most commonly used mechanisms for trace extraction. Ducassé et al. [5] discuss dynamic program analysis and debugging in different logic programming environments. Different approaches are used for extracting an execution trace from a logic program, including source code instrumentation, meta-interpreter instrumentation and compiled code instrumentation.

As the textual representation of the trace is usually not very informative and often difficult to interpret and understand, tools such as JIVE [13] and Java Path Finder [4] support various diagrams built from the execution trace for the easy debugging of Java programs. Maggi et al. [10] introduces the automata-based techniques for the runtime verification of LTL-based process models extracted from the execution logs. In the logic programming context, in addition to domain specific visualization, most logic programming environments support step-by-step execution of programs with limited graphical debugging capability. A good example is SWI-Prolog [12] which provides a step-by-step debugger with views of the call-stack and variable bindings.

The concept of variable binding in constraint languages is more complex than in imperative languages. Carro Liñares et al. [2] presents a visual representation of finite domain CLP programs. This approach mainly focuses on the evolution of variables during the labeling phase of constraint evaluation. The reference [2] also introduces the concept of abstract representation of a variable that has a large number of possible values.

Run-time verification integrates tools and techniques proposed in the formal methods field for the analysis of execution trace [4]. Run-time verification is similar to model checking approach except that the model is built from finite set of traces. RV-Match, RV-Predict and RV-Monitor [4] are some of the examples which use formal analysis methods such as a symbolic execution engine, a semantic debugger, a model checker, and a full-fledged deductive program verifier for the debugging and verification of sequential and object oriented programs. Reference [1] introduces a three-valued semantics to include the meaning of partial observation in run-time verification, where an *inconclusive* decision represents the fact that the trace is not long enough to determine the truth value for temporal specification.

The proposed approach in this paper has much in common with JIVE [13] which supports run-time verification of state diagrams extracted from one or more execution traces. This includes checking consistency of run-time with design-time state diagrams, as well as checking properties stated in CTL. In our approach, the execution trace is implicitly present in the values of series variables at different points in time. In a way, series variables simplify the task of extracting an execution trace.

3 Temporal Constrained Objects

Temporal constrained objects extend the basic paradigm of constrained objects to support time-varying properties of dynamic systems. The TCOB execution follows a discrete time simulation based on the value of built-in variable Time. The user can attain any granularity of time by multiplying a suitable scaling factor with Time, e.g., MyTime = 0.01*Time. The default initial value for Time is equal to 1 unless the different value is specified by the user. A TCOB program defines a collection of classes, each of which contains a set of attributes, constraints, predicates, and constructors [8]. Each of temporal constrained object is an instance of some class whose outline is as follows.

$$
\begin{aligned}
class\_definition &::= [\,\text{abstract}\,]\ \text{class}\ \ class\_id\ \ [\,\text{extends}\ \ class\_id\,]\{\ body\ \} \\
body &::= [\,\text{attributes}\ \ attributes\,] \\
&\quad\ [\,\text{constraints}\ \ constraints\,] \\
&\quad\ [\,\text{predicates}\ \ predicates\,] \\
&\quad\ [\,\text{constructors}\ \ constructors\,]
\end{aligned}
$$

As in Java, single inheritance is defined by the **extends** keyword. An abstract class is a class without a constructor and cannot be instantiated. An attribute is a typed identifier, which support both primitive and user-defined types. The keyword **series** is used to define series variable. The series variable takes on an unbounded sequence of values over time, and temporal constraints are defined in terms of past and future values of the series variable. For every series variable v, the TCOB expression $'v$ and v' refers to the immediate previous and next values of v respectively. These operators can be juxtaposed to refer to successive values of v in the past or future. The past values of v at any point in time are referred to by $'v$, $''v$, $'''v$, ..., and the future values of v at any point in time are referred to by v', v'', v''', ... TCOB also allows a series of variables to be initialized by explicitly assigning values at specific time points. The value of a series variable v at some specific time point i can be accessed by $v<i>$. Appendix A shows the complete grammar of TCOB syntax.

Constraints are relations over the attributes of classes. TCOB supports simple, conditional, quantified and creational constraints over typed attributes. It also provides two *metric temporal operators*, F and G, analogous to 'always' ($\Box$) and 'eventually' ($\Diamond$) operators in (non-metric) temporal logic. In the following, evaluation of F and G operators are carried out with respect to the current value of Time. The constraints specified with metric temporal operator is called metric temporal constraints, which appears in the body part of conditional constraint. The basic meaning of metric temporal constraints are

1. F p states that constraint p must hold at some time point in the future;
2. F<t> p states that constraint p must hold exactly after t units of time; and
3. F<$t1,t2$> p states the constraint p must hold sometime starting after $t1$ units of time but before $t2$ units of time.
4. G p states that constraint p must hold at all time points in the future;
5. G<t> p states that constraint p must hold at all time points after t units of time; and
6. G<$t1,t2$> p states the constraint p must hold at all time points starting after $t1$ units of time but before $t2$ units of time.

Performance is one of the leading considerations for declarative programming languages. We have discussed the performance of TCOB programs is detail in [8], where we have shown the benefits of partial evaluation for improving run-time. This is a compile-time optimization that unravels the time-loop and also eliminates some of the inefficiencies arising due to layers of predicate calls and also due to repeated consistency checks of the constraint store during constraint-solving. By generating a reduced set of constraints as its output, partial evaluation also helps make our language independent of specific constraint solvers. The following section explains the PID controller and traffic lights example from [8].

PID Controller Example: A *Proportional Integral Derivative* (PID) controller is one of the most commonly used control-loop feedback mechanism in industrial automation. The PID controller consists of three components: controller, plant, and sensor [11]. The output of the controller is given as the control input to the plant. The sensor collects the output from the plant and calculates the error based on the desired output. The estimated error is fed back as the controller input. The controller is defined by the equation.

$$u(t) = K_p e(t) + K_i \int_0^t e(t)dt + K_d \frac{de(t)}{dt}$$

Here $u(t)$ represents the output from the controller; $e(t)$ is the error feedback to the controller; and the K_p, K_i and K_d are the non-negative coefficients. The first term denotes the present value of the error; the second term indicates the past values of error; and the last term accounts for future values of error. The corresponding TCOB formulation is given below:

```
class controller {
  attributes
     int Kp, Ki, Kd;
     series real Error,ESum,Out;
  constraints
     Kp > 0; Kd > 0; Ki > 0;
     ESum = Error + 'ESum;
     Out = (Kp * Error) + (Ki * ESum) + Kd * (Error - 'Error);
  constructors controller(KP, KI, KD) {
     Kp = KP; Ki = KI; Kd = KD;
     Error<1> = 0; ESum<1> = 0; Error<2> = 0; ESum<2> = 0;
  }
}
```

The series variables Error and Out define the input and output of the controller respectively. The constraint section models the PID controller equation. For the **plant** model, we consider a simple mass spring damper problem. The modeling equation is

$$M\frac{d^2x}{d^2t} + b\frac{dx}{dt} + kx = F$$

where x is the displacement of body, M represents the mass of the body, and b and k are the damping constant and spring constant, and F is the force applied on the body to position it. In each iteration, the controller calculates a new force and the plant takes it as the input. In the equilibrium condition of this experiment, the displacement value remains at one.

```
class plant{
  attributes
     series real Fo, X, V;
     int M,B,K;
  constraints
     V = X - 'X;
     Fo = M * (V - 'V)+ B * V + K * X;
  constructor plant(M1,B1,K1,X1,V1)
     { M = M1; B = B1; K = K1; X<1> = X1; V<1> = V1; }
}
```

The sensor takes the output from the plant, and the system works out the error and feedback to the controller. Appendix B gives the definition of **sensor** and **system** class.

The error is the difference between the expected value (here the displacement $= 1$) and sensed value. In our experiments, we modeled the system by assigning the value one to the coefficients K_p, K_i and K_d. We can fine tune this value based on the system behavior.

Traffic Light Example: Consider two traffic lights which control respectively the traffic in east-west and north-south intersections by sensing the traffic on these roads and controlling the duration of the green and red lights depending upon the number of vehicles on each road. We refer to these as "intelligent" traffic lights and they are modeled by class int_light below. The traffic_sensor class generates random values (between 0 and 5) to represents the number of new vehicles that have arrived at every time point. The int_light class sums up these values to determine the total number of vehicles waiting while the light is not green. When the light changes to green, the number of waiting vehicles determines the duration of the green light (5 time steps for each vehicle); if this number is zero, the green light is skipped. The remaining constraints specify the safety and synchronization properties between traffic lights.

```
class int_light{
  attributes
      enum  Color = [red,green,yellow];
      series Color C;
      traffic_sensor Ts; int_light Tl;
      series int NumV;
  constraints
      C = green --> NumV = 0;
      not ( C = green ) --> NumV = 'NumV + Ts.V;
      'C = red & C = yellow & NumV > 0 -->
              G<1,NumV * 5> C = green & F<NumV * 5> C = yellow;
      'C = green & C = yellow & Tl.NumV = 0 -->
              G<1,NumV * 5> C = green & F<NumV * 5> C = yellow;
      Tl.C = yellow --> C = yellow; Tl.C = green --> C = red;
  constructors int_light(C1,C2,TL){
      Tl = TL; C<1> = C1; C<2> = C2; NumV<2> = Ts.V<2>;
      Ts = new traffic_sensor();
  }
}
class system {
  attributes
      int_light NS, EW;
  constructor system(){
      NS = new int_light(red,yellow,EW);
      EW = new int_light(green,yellow,NS);
  }
}
```

The system class initializes the two traffic lights with initial colors. The constructor call NS = new int_light(red,yellow,EW) creates the North-South traffic light with initial colors and traffic light in the opposite direction. The NS object makes use of EW object attributes for color synchronization. Note that EW is unbound when NS is initialized, but the implementation of conditional constraints handles this situation by keeping the unbounded object constraints in the constraint store and automatically invoking it when EW becomes bound.

4   Run-Time Analysis

Run-time analysis of temporal constrained objects consists of three main components: visualization, error detection, and verification. Figure 1 gives an overview of these three aspects, which are all driven by an execution trace. In order to obtain this execution trace, the body of every class definition may include an optional 'monitor' clause which specifies which class attributes are to be monitored during execution. Here, *attributes* is a comma-separated list of attribute names, and they can be both series variables as well as non-series variables. The following example illustrate the use of 'monitor' clause in a simple program of a moving object whose position at each step depends on the previous position and a constant value.

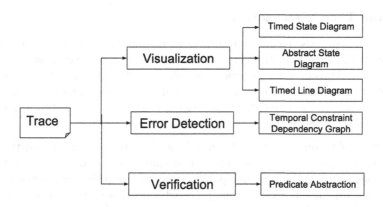

Fig. 1. Overview of run-time analysis

```
class example {
  attributes
    series int P; int C;
  constraints
    P > 0;
    P = 'P + C;
  constructors example()
    C = -1;   P<1> = 2;
  monitor P,C;
}
```

At run-time, the underlying system creates an execution trace file containing, for every time-point of execution, the attribute name, object reference and value for every designated attribute in every monitor clause in the program. The following is a sample trace file for the above example, here we have added a monitor clause for attribute P and C in class plant.

```
Time = 2, Obj = P, Var = P, Val = NaV
Time = 2, Obj = P, Var = X, Val = 1
Time = 2, Obj = C, Var = X, Val = -1
Time = 3, Obj = P, Var = X, Val = NaV
Time = 3, Obj = C, Var = X, Val = -1
.......
```

The monitor clause effectively causes a listener to be placed on each attribute to be monitored. The monitor outputs a value 'NaV' (Not a Value) if the attribute is undefined. The trace file may contains multiple value for same attributes at each instance of time due to the complex backtracking in the underlying computing engine.

4.1   Run-Time Visualization

Visualization conveys information in a more readable and efficient way and, in our approach, it serves as the foundation for debugging and verification.

Our analysis framework provides a web-based GUI (Fig. 2a), where the user can upload an execution trace file generated by the monitored program and then select one or more attributes for visualization. Currently, the framework supports three types of diagrams, as described below.

Timed State Diagram. A state diagram is a more precise way to portray the evolution of a system over time. The timed state diagram describes the values of variables at each discrete time-points. Once a non-series variable is bound, its value remains fixed at every time-point whereas series variables may assume different values at different time-points. For example, Fig. 2b shows the timed state diagram of a traffic light example.

Abstract State Diagram. The abstract state digram shows the abstract view of the system by extracting the distinct states of the execution. The timed state diagram shows the linear progression of the system with respect to time. This diagram is not an appropriate choice when we wish to visualize the repetitive behavior of a system. The following procedure can build the abstract state diagram from a trace by considering the state vector in each trace entry.

```
Trace T;
Abstract_State_Graph AS = {}
foreach E<time,state> in T
  if (E.state not in AS)
     AS = AS U {E.state};
```

Figure 2c present the abstract view of traffic light example, whereas the equivalent timed state diagram would contain as many states as the number of simulation steps.

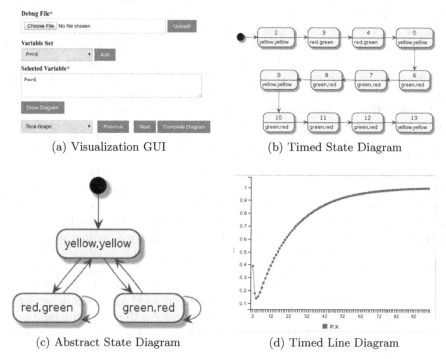

(a) Visualization GUI   (b) Timed State Diagram

(c) Abstract State Diagram   (d) Timed Line Diagram

Fig. 2. Run-time visualization of TCOB programs

Timed Line Diagram. The timed line diagram plots the values of the chosen variables over time. Unlike other diagrams, it supports only numerical attributes but is often useful in identifying incorrectness by direct inspection of the form the output diagram. For example, the Fig. 2d shows the timed line diagram for displacement(X) in the PID controller example. It captures the correctness of PID controller implementation, since the displacement should eventually reach a stable state.

4.2 Run-Time Error Detection

Several factors contribute to an erroneous output of a TCOB program and it is hard to pinpoint the cause of failure in a complex large-scale simulation. Here we illustrate the various methods to identify the cause of errors in TCOB simulation using run-time analysis.

Case 1: Constraint Failure. The execution of a TCOB program involves a discrete-time simulation with constraint satisfaction performed at each time-point, where the constraints may involve non-series as well as series variables. Each time-step can be viewed as a *computation frame* that involves values of series variables from next and previous time-steps. The size of the computation frame depends on the next and previous reference of series variables in the constraints; the default size is 1 when there are no such references.

Consider the position example described above. The programmer erroneously enters a negative value (-1) for the constant. As a result, the unary constraint P>0 fails at Time = 3 since the value of the series variable P is 2 at Time = 1 and it decreases by 1 at each time-step.

In general, the interweaving of constraint solving within a time-based simulation makes it is difficult to detect this type of programming error in a larger system. In order to address this problem, we propose the use of a *temporal constraint dependency graph* using the computation frame of TCOB execution. In this example, the size of computation frame is 2. Each computation frame maintains two sets: bound variables (BV) and unbound variables (UV). The bound variables set maintains the set of ground variables along with their values. Both sets are computed before constraint solving is initiated at each time step. Figure 3a gives the computation frame and temporal constraint dependency graph at Time = 2. The edge between nodes indicates a (binary) constraint relation and the self-loop indicate the unary constraints. The vertical dotted lines delineate the computation frame.

The computation frame moves by one time unit at each execution step. For example, at Time = 3 the computation frame is shown in Fig. 3b. The dotted edge indicates the relation in previous computation frame. During the constraint solving at Time = 3, the variable P<3> is assigned to 0 and causes a violation of the unary constraint P>0. The user can quickly backtrack and find which assignment or constraint leads to this inconsistency. The red edge shows the constraint violation in the system.

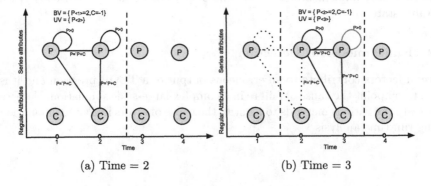

(a) Time = 2   (b) Time = 3

Fig. 3. Temporal constraint dependency graph (Color figure online)

Case 2: Incorrect and Undefined Answers. Incorrectness arises when constraint satisfaction results in a successful outcome but the computed answer for one or more variables is incorrect. The probable cause could be incorrect constraints, erroneous initialization or possibly incorrect assembly of objects. The user must have enough domain knowledge to distinguish which of these cases is the real cause. When the programmer has some partial knowledge about the output, s/he can encode it as a constraint in the program.

One of the primary advantages of programming with constraints is their ability to compute with partial (or incomplete) information. Sometimes partial information is not adequate to compute a specific value for a variable. This usually happens in TCOB due to either missing variable initialization or missing constraints. Consider the following constructor from the `controller` class in the PID controller example.

```
controller(KP,KI,KD){
      Kp = KP; Ki = KI; Kd = KD;
      Error<2> = 0; Error<1> = 0;
}
```

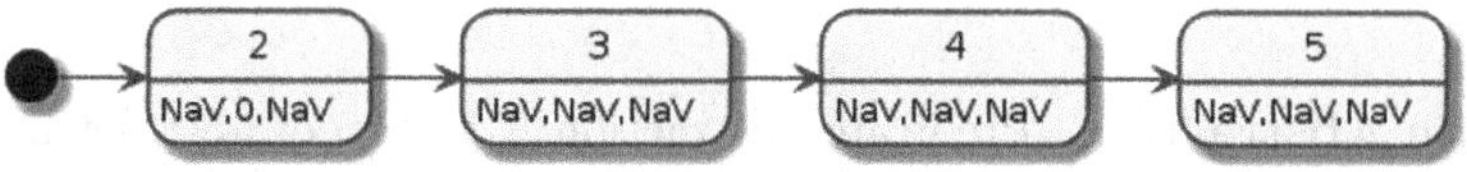

Fig. 4. Time based state diagram of TCOB execution

Suppose we monitor the series variables `ESum`, `Error` and `Out`. Figure 4 presents the state diagram of execution, which is built from the execution trace of the PID controller program. The figure clearly shows that the constraints failed to determine exact values for `ESum` and `Out` using the available information at time step 2. The current value `ESum` depends on `Error` and previous value of `ESum`. The initialization `ESum<1> = 0` in the constructor can correct this problem.

4.3   Run-Time Verification

We now illustrate the concept of run-time verification for temporal constrained objects. In a model-based approach [3], we check the consistency of a model M against a specification ϕ. In run-time verification, execution traces are used to build a run-time model and properties are verified for this model. Both timed as well as the abstract state diagram can serve as the basis of a model. However, the abstract state diagram is a more compact view of the timed state diagram; it is essentially a (run-time) Kripke structure [3] in the terminology the model checking. The safety and liveness verification conditions are specified as LTL formula; if an LTL condition is true in the abstract state diagram it is also true in the timed state diagram.

For example, the abstract state diagram for the traffic light example of Sect. 3 is given in Fig. 2c. In this example, safety means that two lights are not green at the same time, and liveness means that the lights always changing without being stuck at any state. The corresponding LTL formulation is as follows, where NS and EW refer to north-south and east-west respectively. From Fig. 2c, it is evident that all these conditions are satisfied by our model.

$$\Box\neg((NS.C = green) \wedge (EW.C = green))$$
$$\Box(NS.C = green \implies \Diamond NS.C = red)$$
$$\Box(NS.C = red \implies \Diamond NS.C = green)$$
$$\Box(EW.C = green \implies \Diamond EW.C = red)$$
$$\Box(EW.C = red \implies \Diamond EW.C = green)$$

There is an important difference between a run-time state diagram derived from a finite execution trace and the design-time state diagrams used in model-checking. Whereas cycles in the design-time state diagrams represent nonterminating execution paths, cycles in run-time state diagrams represent finite execution paths. For example, in Fig. 2c, the self-loops are executed only a finite number of time-steps.

Predicate Abstraction. Predicate abstraction is a form of abstract interpretation wherein we can reduce the size of the model constructed by abstracting details [6]. For example, if in some analysis we care only whether an integer variable is negative, zero, or positive, we can effectively abstract the infinite set of integers by a set with just three values. TCOB simulations can result in a large number of states causing a state explosion problem during run-time verification. Predicate abstraction is very useful in reducing the run-time state diagrams. This abstraction can directly specified as declarative constraints in a TCOB class definition.

For example, consider the series variable X from the **plant** class in PID controller example. This models the displacement, a real number, which has minute differences from one time-step to another. We need to verify this variable X eventually reaches a stable state, where value is approximately equal to required displacement value, namely, 1.

In such scenarios, we can use predicate abstraction by grouping the displacement value into in a few different ranges using constraints, as shown below.

```
X > 0.4 & X < 0.8 --> PV = 0.5;
X < 0.4 --> PV = 0.0;
X > 0.8 & X < 1.2 --> PV = 1;
X > 1.2 --> PV = 1.5
```

Here PV is a series variable. For a 1000-step simulation, there would be 1000 different values X, but predicate abstraction reduces them to three values and hence three states in the state diagram. The state diagram would uphold the LTL condition: $\Box(\Diamond PV = 1)$. This formula represents the stability of PID controller by ensuring the displacement reaches the user-specified value.

Run-time verification is aimed at analyzing individual executions of a system. In comparison with traditional verification, which is more complex, run-time verification is simpler but has the shortcoming of sacrificing full coverage. However, many applications that are modeled with state diagrams involve a repetitive cycle and a single run often covers many scenarios. And, through the availability of a good set of test cases, coverage of all possible behaviors can be better approximated.

5   Conclusions and Further Work

The main contribution of this paper is a set of techniques for run-time analysis of temporal constrained objects. The execution of temporal constrained objects involves a time-based simulation together with constraint solving at each time-step, where the constraints could involve ordinary variables as well as series variables, which may take different values at each time-step. We have developed techniques for error detection as well as reasoning about the correctness of execution, i.e., run-time verification. In the former case, we make use a temporal constraint dependency graph and in the latter case, we construct finite state machines which serve as a basis for (run-time) model-checking using propositional temporal logic. In both cases, we make crucial use of the execution trace of the program.

The ideas discussed in this paper have been implemented as part of the TCOB language and execution environment. The TCOB compiler translates TCOB programs to SWI-Prolog programs. Run-time analysis is carried out starting from the execution trace that is generated based upon the monitor clauses in the program. Temporal constrained objects simplify the task of constructing an execution trace, because this is implicitly present in the values bound to the series variables. The state diagrams were constructed using the PlantUML drawing package. As part of our future work, we propose to apply our run-time analysis methodology to larger applications and combine execution, visualization, error analysis and run-time verification in more integrated manner.

Appendix

A. TCOB Grammar

$$
\begin{aligned}
class\_definition \ &::= [\ \text{abstract}\ ]\ \text{class}\ class\_id\ [\ \text{extends}\ class\_id\ ]\{\ body\ \} \\
body \ &::= [\ \text{attributes}\ attributes\ ] \\
&\quad [\ \text{constraints}\ constraints\ ] \\
&\quad [\ \text{predicates}\ predicates\ ] \\
&\quad [\ \text{constructors}\ constructors\ ] \\
attributes \ &::= [\ decl\ ;\ ]^+ \\
decl \ &::= [\text{series}]\ type\ id\_list \\
type \ &::= primitive\ |\ class\_id\ |\ type[\ ] \\
primitive \ &::= \text{real}\ |\ \text{int}\ |\ \text{bool}\ |\ \text{char}\ |\ \text{string} \\
id\_list \ &::= attribute\_id\ [\ ,\ attribute\_id\ ]^+ \\
constraints \ &::= [\ constraint\ ;]^+ \\
constraint \ &::= creational\ |\ quantified\ |\ simple \\
creational \ &::= attribute\ = \text{new}\ class\_id(terms) \\
quantified \ &::= \text{forall}\ var\ \text{in}\ enum\ :\ constraint\ | \\
&\quad \text{exists}\ var\ \text{in}\ enum\ :\ constraint \\
simple \ &::= conditional\ |\ constraint\_atom \\
conditional \ &::= literals --> condi\_body \\
condi\_body \ &::= mto\_literals\ [\&\ mto\_literals]^+ \\
mto\_literals \ &::= literal\ |\ mto\_constraint \\
mto\_constraint \ &::= \text{F}\ constraint\_atom\ |\ \text{F} < interval >\ constraint\_atom\ | \\
&\quad \text{F} < interval, interval >\ constraint\_atom\ |\ \text{G}\ constraint\_atom\ | \\
&\quad \text{G} < interval >\ constraint\_atom\ | \\
&\quad \text{G} < interval, interval >\ constraint\_atom \\
constraint\_atom \ &::= term\ relop\ term\ |\ cpred\_id(terms) \\
relop \ &::= =\ |\ !=\ |\ >\ |\ <\ |\ >=\ |\ <=
\end{aligned}
$$

B. PID Controller

```
class controller {
 attributes
   int Kp, Ki, Kd;
   series real Error,ESum,Out;
 constraints
   Kp > 0; Kd > 0; Ki > 0;
   ESum = Error + 'ESum;
   Out = Kp * Error + Ki * ESum
       + Kd * (Error - 'Error);
 constructors
   controller(KP, KI, KD) {
     Kp = KP; Ki = KI; Kd = KD;
     Error<1> = 0; ESum<1> = 0;
     Error<2> = 0; ESum<2> = 0;
   }
}

class plant {
 attributes
   series real Fo, X, V;
   int M,B,K;
 constraints
   V = X-'X;
   Fo = M *(V-'V)+ B * V + K*X;
 constructor
   plant(M1,B1,K1,X1,V1){
     M = M1; B = B1; K = K1;
     X<1> = X1; V<1> = V1;
   }
}
```

```
class sensor {
 attributes
   series real Output;
   plant P;
 constraints
   P.X = Output;
 constructor sensor(P1)
   { P = P1;}
}

class system {
 attributes
   plant P; controller C;
   sensor S; real Dvalue;
 constraints
   P.Fo = C.Out;
   Dvalue = C.Error' + S.Output;
 constructor system() {
   P = new plant(1,10,20,0,1);
   C = new controller(1,1,1);
   S = new sensor(P);
   Dvalue = 1;
 }
}
```

References

1. Bauer, A., Leucker, M., Schallhart, C.: Runtime verification for LTL and TLTL. ACM Trans. Softw. Eng. Methodol. (TOSEM) **20**(4), 14 (2011)
2. Carro Liñares, M., Hermenegildo, M.V.: Visualization designs for constraint logic programming. Upgrade **2**(2), 27–34 (2001)
3. Clarke, E.M., Grumberg, O., Peled, D.: Model Checking. MIT press, Cambridge (1999)
4. Daian, P., et al.: Runtime verification at work: a tutorial. In: Falcone, Y., Sánchez, C. (eds.) RV 2016. LNCS, vol. 10012, pp. 46–67. Springer, Cham (2016). https://doi.org/10.1007/978-3-319-46982-9_5
5. Ducassé, M., Noyé, J.: Logic programming environments: dynamic program analysis and debugging. J. Log. Program. **19**, 351–384 (1994)
6. Flanagan, C., Qadeer, S.: Predicate abstraction for software verification. In: ACM SIGPLAN Notices, vol. 37, pp. 191–202. ACM (2002)

7. Jayaraman, B., Tambay, P.: Modeling engineering structures with constrained objects. In: Krishnamurthi, S., Ramakrishnan, C.R. (eds.) PADL 2002. LNCS, vol. 2257, pp. 28–46. Springer, Heidelberg (2002). https://doi.org/10.1007/3-540-45587-6_4

8. Kannimoola, J.M., Jayaraman, B., Tambay, P., Achuthan, K.: Temporal constrained objects: application and implementation. Comput. Lang. Syst. Struct. **49**, 82–100 (2017)

9. Leucker, M., Schallhart, C.: A brief account of runtime verification. J. Log. Algebr. Program. **78**(5), 293–303 (2009)

10. Maggi, F.M., Westergaard, M., Montali, M., van der Aalst, W.M.P.: Runtime verification of LTL-based declarative process models. In: Khurshid, S., Sen, K. (eds.) RV 2011. LNCS, vol. 7186, pp. 131–146. Springer, Heidelberg (2012). https://doi.org/10.1007/978-3-642-29860-8_11

11. Matlab: Introduction: PID Controller Design (2015). http://ctms.engin.umich.edu/CTMS/index.php?example=Introduction&section=ControlPID

12. Wielemaker, J., Schrijvers, T., Triska, M., Lager, T.: SWI-prolog. Theory Pract. Log. Program. **12**(1–2), 67–96 (2012)

13. Ziarek, L., Jayaraman, B., Lessa, D., Swaminathan, J.: Runtime visualization and verification in JIVE. In: Falcone, Y., Sánchez, C. (eds.) RV 2016. LNCS, vol. 10012, pp. 493–497. Springer, Cham (2016). https://doi.org/10.1007/978-3-319-46982-9_33

Implementation of Logical Retraction in Constraint Handling Rules with Justifications

Thom Frühwirth[(✉)]

Ulm University, Ulm, Germany
thom.fruehwirth@uni-ulm.de

Abstract. In previous work we added justifications to Constraint Handling Rules (CHR) to enable logical retraction of constraints for dynamic algorithms. We presented a straightforward source-to-source transformation to implement this conservative extension. In this companion paper, we improve the performance of the transformation. We discuss its worst-case time complexity in general. Then we perform experiments. We benchmark the dynamic problem of maintaining shortest paths under addition and retraction of paths. The results validate our complexity considerations.

1 Introduction

Justifications have their origin in truth maintenance systems (TMS) [McA90] for automated reasoning. Derived information (a formula) is explicitly stored and associated with the information it originates from by means of justifications. With the help of justifications, conclusions can be withdrawn (undone) by retracting their premises. By this *logical retraction*, inconsistencies can be repaired by retracting one of the reasons for the inconsistency.

In the formalism and programming language Constraint Handling Rules (CHR) [Frü09, Frü15, FR18], conjunctions of atomic formulae (constraints) are rewritten by rule applications. When algorithms are written in CHR, constraints represent both data and operations. CHR is already incremental by nature, i.e. constraints can be added at runtime. Logical retraction then adds decrementality. To accomplish logical retraction in CHR, we have to be aware that CHR constraints can also be deleted by rule applications. These constraints may have to be restored when a premise constraint is retracted. With logical retraction, any algorithm written in CHR becomes *fully dynamic*[1]. Operations can be undone and data can be removed at any point in the computation without compromising the correctness of the result.

In [Fru17], we formally defined a correct conservative extension of CHR with justifications (CHR$^\mathcal{J}$). We gave a straightforward source-to-source transformation that adds justifications for user-defined constraints. A scheme of two rules

---

[1] Dynamic algorithms for dynamic problems should not be confused with dynamic programming.

© Springer Nature Switzerland AG 2018
D. Seipel et al. (Eds.): DECLARE 2017, LNAI 10997, pp. 37–52, 2018.
https://doi.org/10.1007/978-3-030-00801-7_3

sufficed to allow for logical retraction (deletion, removal) of constraints during computation. Without the need to recompute from scratch, these rules retract not only the constraint but also undo all consequences of the rule applications that involved the constraint.

The runtime performance of the previous translation scheme is not optimal, however. In this paper, we present an improved source-to-source transformation for logical retraction of constraints with justifications in CHR (CHR$^{\mathcal{J}}$). This transformation only imposes a constant factor overhead as long as justifications are not used for retraction. We will argue that the worst-case time complexity for any number of retractions is in general proportional to the number of rule applications, i.e. derivation length. The complexity of an algorithm expressed in CHR is usually a polynomial in the derivation length. Therefore retraction indeed has typically less complexity than recomputation from scratch at the expense of storing removed constraints. The added space complexity is again bounded by the derivation length. In our experiments, we will consider the dynamic problem of maintaining shortest paths under addition and retraction of paths.

Minimum Example. Given a multiset of numbers represented as conjunction $\text{min}(n_1),\text{min}(n_2),\ldots,\text{min}(n_k)$. The constraint (predicate) $\text{min}(n_i)$ means that the number n_i is a candidate for the minimum value. The following CHR rule filters the candidates.

```
min(N) \ min(M) <=> N=<M | true.
```

The rule consists of a left-hand side, on which a pair of constraints has to be matched, a guard check N=<M that has to be satisfied, and an empty right-hand side denoted by `true`. In effect, the rule takes two `min` candidates and removes the one with the larger value (constraints after the \ symbol are deleted). Note that the `min` constraints behave both as operations (removing other constraints) and as data (being removed).

CHR rules are applied exhaustively. Here the rule keeps on going until only one, thus the smallest value, remains as single `min` constraint, denoting the current minimum. If another `min` constraint is added during the computation, it will eventually react with a previous *min* constraint, and the correct current minimum will be computed in the end. Thus the algorithm as implemented in CHR is incremental. It is not decremental, though: We cannot logically retract a *min* candidate. While removing a candidate that is larger than the minimum would be trivial, the retraction of the minimum itself requires to remember all deleted candidates and to find their minimum. As we will see, with the help of justifications, this logical retraction will be possible automatically.

Related Work. The work of Armin Wolf on Adaptive CHR [WGG00] introduced justifications into CHR. Different to our work, this technically involved approach requires to store detailed information about the rule instances that have been applied in a derivation in order to undo them. In our approach, we use a straightforward source-to-source transformation and retract constraints one-by-one instead. Adaptive CHR had a low-level implementation in Java [Wol01], while we give an implementation in CHR itself by source-to-source transformations. The more recent work of Duck [Duc12] introduces SMCHR, a tight integration of CHR with a Boolean Satisfiability (SAT) solver for quantifier-free formulae including disjunction and negation as logical connectives. It is mentioned without giving further details that for clause generation, SMCHR supports justifications.

Overview of the Paper. In the next section we recall abstract syntax and refined operational semantics for CHR. In Sect. 3, we describe CHR with justifications for logical retraction of constraints and its previous implementation by a straightforward source-to-source transformation. In Sect. 4, our current work is to optimize this implementation and to discuss its worst-case run-time complexity. In Sect. 5, we report on the results of experiments with our new implementation for the dynamic problem of maintaining shortest paths in a graph under addition (insertion) and deletion (retraction) of paths. The paper ends with conclusions and directions for future work.

2   Preliminaries

We recall abstract syntax and refined operational semantics of CHR [Frü09] in this section.

2.1   Abstract Syntax of CHR

Constraints are relations, distinguished predicates of first-order predicate logic. We differentiate between two kinds of constraints: *built-in (pre-defined) constraints* and *user-defined (CHR) constraints* which are defined by the rules in a CHR program.

Definition 1. A *CHR program* is a finite set of rules. A *(generalized) simpagation rule* is of the form

$$r : H_1 \backslash H_2 \Leftrightarrow C | B$$

where r: is an optional *name* (a unique identifier) of a rule. In the rule *head* (left-hand side), H_1 and H_2 are conjunctions of user-defined constraints, the optional *guard* C is a conjunction of built-in constraints, and the *body* (right-hand side) B is a goal. A *goal* is a conjunction of built-in and user-defined constraints. A *state* is a goal. Conjunctions are understood as *multisets* of their conjuncts.

In the rule, H_1 are called the *kept constraints*, while H_2 are called the *removed constraints*. At least one of H_1 and H_2 must be non-empty. If H_1 is empty, the rule corresponds to a *simplification rule*, also written

$$s : H_2 \Leftrightarrow C | B.$$

If H_2 is empty, the rule corresponds to a *propagation rule*, also written

$$p : H_1 \Rightarrow C | B.$$

In this work, we restrict given CHR programs to rules without built-in constraints in the body except *true* and *false*.

2.2   Operational Semantics of CHR

We follow the exposition in [SF06] in this subsection. Given a query, the rules of the program are applied to exhaustion. A rule is applicable, if its head constraints are matched by constraints in the current goal one-by-one and if, under this matching, the guard check of the rule holds. More formally, the guard is logically implied by the built-in constraints in the goal.

Any of the applicable rules can be applied, and the application cannot be undone, it is committed-choice (in contrast to Prolog). When a simplification rule is applied, the matched constraints in the current goal are replaced by the body of the rule, when a propagation rule is applied, the body of the rule is added to the goal without removing any constraints. When a simpagation rule is applied, only the head constraints right to the backslash symbol are removed, the head constraints before are kept.

As in Prolog, almost all CHR implementations execute queries from left to right and apply rules top-down in the textual order of the program. This behavior has been formalized in the so-called refined semantics that was also proven to be a concretization of the standard operational semantics [DSdlBH04]. In this refined semantics of actual implementations, a CHR constraint in a query can be understood as a procedure that goes efficiently through the rules of the program in the order they are written.

We consider such a constraint to be *active*. When it matches a head constraint of a rule, it will look for the other, *partner constraints* of the head in the *constraint store* and check the guard until an applicable rule is found. If the active constraint has not been removed after trying all rules, it will be delayed and put into the constraint store as data. Constraints from the store will be reconsidered (woken) if newly added built-in constraints constrain variables of the constraint, because then rules may become applicable since their guards are now implied.

Hash Indexing in CHR. For optimal time complexity, (near) constant-time addition, finding and removal of CHR constraints is required. To achieve this efficiency, CHR implementations typically provide for indexing on arguments of constraints. Most current CHR libraries in Prolog are based on the KU Leuven CHR system. It supports indexing for terms via attributed variables, in SWI Prolog also hash tables for ground terms and arrays for dense integers.

The hash table based indexes in SWI Prolog work at the argument level. For efficient constraint lookups, these arguments have to be ground during computation. For optimal generation of indexes, the SWI Prolog CHR system depends on mode and type information specified in constraint declarations. We will therefore use these declarations for justifications and other constraint arguments in our implementation examples.

3   CHR with Justifications (CHR$^{\mathcal{J}}$)

We present a conservative extension of CHR by justifications following [Fru17]. If justifications are not used, programs behave as without them. Justifications annotate atomic CHR constraints. A straightforward source-to-source transformation extends the rules with justifications.

3.1 CHR with Justifications for Logical Retraction

We start with adding justifications to CHR constraints and states.

Definition 2 (CHR Constraints and Initial States with Justifications).
A *justification* f is a unique identifier. Given an atomic CHR constraint G, a *CHR constraint with justifications* is of the form G^F, where F is a set of justifications. An *initial state with justifications* is of the form $\bigwedge_{i=1}^{n} G_i^{\{f_i\}}$ where the f_i are distinct justifications.

We now define a source-to-source translation from rules to rules with justifications. Let *kill* (retract) and *rem* (remember removed) be to unary *reserved* CHR constraint symbols. This means they are only allowed to occur in rules as specified in the following.

Definition 3 (Translation to Rules with Justifications). Given a generalized simpagation rule

$$r : \bigwedge_{i=1}^{l} K_i \ \backslash \ \bigwedge_{j=1}^{m} R_j \Leftrightarrow C \mid \bigwedge_{k=1}^{n} B_k$$

Its translation to a *simpagation rule with justifications* is of the form

$$rf : \bigwedge_{i=1}^{l} K_i^{F_i} \ \backslash \ \bigwedge_{j=1}^{m} R_j^{F_j} \Leftrightarrow C \mid \bigwedge_{j=1}^{m} rem(R_j^{F_j})^F \wedge \bigwedge_{k=1}^{n} B_k^F \text{ where } F = \bigcup_{i=1}^{l} F_i \cup \bigcup_{j=1}^{m} F_j.$$

The translation ensures that the head and the body of a rule mention exactly the same justifications. The reserved CHR constraint *rem/1* (remember removed) stores the constraints removed by the rule together with their justifications.

Translating the minimum rule from the introduction to one with justifications results in:

$$min(A)^{F_2} \ \backslash \ min(C)^{F_2} \Leftrightarrow A < C \mid F = F_1 \cup F_2 \wedge rem(min(C)^{F_2})^F.$$

To avoid clutter, let $A^{\mathcal{J}}, B^{\mathcal{J}}, C^{\mathcal{J}} \ldots$ stand for conjunctions (or corresponding states) whose atomic CHR constraints are annotated with justifications according to the above definition of the rule scheme. Similarly, let $rem(R)^{\mathcal{J}}$ denote a conjunction $\bigwedge_{j=1}^{m} rem(R_j^{F_j})^F$.

We showed previously that rule applications correspond to each other in standard CHR and in CHR$^{\mathcal{J}}$.

Lemma 1 (Equivalence of Program Rules). [Fru17] There is a computation step $S \mapsto_r T$ with simpagation rule

$$r : H_1 \backslash H_2 \Leftrightarrow C \mid B$$

if and only if there is a computation step with justifications $S^{\mathcal{J}} \mapsto_{rf} T^{\mathcal{J}} \wedge rem(H_2)^{\mathcal{J}}$ with the corresponding simpagation rule with justifications

$$rf : H_1^{\mathcal{J}} \backslash H_2^{\mathcal{J}} \Leftrightarrow C | rem(H_2)^{\mathcal{J}} \wedge B^{\mathcal{J}}.$$

Since computations are sequences of connected computation steps, this lemma implies that computations in standard CHR program and in $CHR^{\mathcal{J}}$ correspond to each other. Thus CHR with justifications is a conservative extension of CHR.

Logical Retraction Using Justifications. We use justifications to retract a CHR constraint from a computation without the need to recompute from scratch. This means that all its consequences due to rule applications it was involved in are undone. CHR constraints added by those rules are removed and CHR constraints removed by the rules are re-added (inserted). To specify and implement this behavior, we give a scheme of two rules, one for retraction and one for re-adding of constraints. The reserved CHR constraint $kill(f)$ (retract) undoes all consequences of the constraint with justification f.

Definition 4 (Rules for CHR Logical Retraction). For each n-ary CHR constraint symbol c (except the reserved *kill* and *rem*), we add a rule to kill constraints and a rule to revive removed constraints of the form:

$$\text{kill} : kill(f) \backslash G^F \Leftrightarrow f \in F \mid true$$

$$\text{revive} : kill(f) \backslash rem(G^{F_c})^F \Leftrightarrow f \in F \mid G^{F_c},$$

where $G = c(X_1, \ldots, X_n)$, where $X_1, \ldots, X_n$ are different variables.

Note that a constraint may be revived and subsequently killed. This is the case when both F_c and F contain the justification f.

We proved previously correctness of logical retraction: the result of a computation with retraction is the same as if the constraint would never have been introduced in the computation. We showed that given a computation starting from an initial state with a $kill(f)$ constraint that ends in a state where the *kill* and *revive* rules have been applied to exhaustion, then there is a corresponding computation without constraints that contain the justification f.

Theorem 1 (Correctness of Logical Retraction). [Fru17] Given a computation

$$A^{\mathcal{J}} \wedge G^{\{f\}} \wedge kill(f) \mapsto^* B^{\mathcal{J}} \wedge rem(R)^{\mathcal{J}} \wedge kill(f) \not\mapsto_{kill,revive,}$$

where f does not occur in $A^{\mathcal{J}}$. Then there is a computation without $G^{\{f\}}$ and $kill(f)$

$$A^{\mathcal{J}} \mapsto^* B^{\mathcal{J}} \wedge rem(R)^{\mathcal{J}}.$$

3.2   Previous Implementation

We recall the implementation of [Fru17] for CHR with justifications ($CHR^{\mathcal{J}}$).

Constraints with Justifications. CHR constraints annotated by a set of justifications are realized by a binary infix operator ##, where the second argument is a list of justifications:

$C^{\{F_1,F_2,\dots\}}$ is realized as C ## [F1,F2,...].

For convenience, we add rules that add a new justification to a given constraint C. For each constraint symbol c with arity n there is a rule of the form

addjust @ c(X1,X2,...Xn) <=> c(X1,X2,...Xn) ## [_F].

where the arguments of X1,X2,...Xn are different variables.

Rules with Justifications. A CHR simpagation rule with justifications is realized as follows:

$$rf : \bigwedge_{i=1}^{l} K_i^{F_i} \setminus \bigwedge_{j=1}^{m} R_j^{F_j} \Leftrightarrow C \mid \bigwedge_{j=1}^{m} rem(R_j^{F_j})^F \wedge \bigwedge_{k=1}^{n} B_k^F \text{ where } F = \bigcup_{i=1}^{l} F_i \cup \bigcup_{j=1}^{m} F_j$$

```
rf @ K1 ## FK1,... \ R1 ## FR1,... <=> C |
    union([FK1,...FR1,...],Fs), rem(R1##FR1) ## Fs,...B1 ## Fs,...
```

where the auxiliary predicate union/2 computes the ordered duplicate-free union of a list of lists[2].

Rules kill, remove *and* revive. Justifications are realized as *flags* that are initially unbound logical variables. This eases the generation of new unique justifications and their use in killing. Concretely, the reserved constraint $kill(f)$ is realized as built-in equality F=r, i.e. the justification variable gets bound. If $kill(f)$ occurred in the head of a *kill* or *revive* rule, it is moved to the guard as equality test F==r.

revive : $kill(f) \setminus rem(C^{F_c})^F \Leftrightarrow f \in F \mid C^{F_c}$
kill : $kill(f) \setminus C^F \Leftrightarrow f \in F \mid true$

```
revive @ rem(C##FC) ## Fs <=> member(F,Fs),F==r | C ## FC.
remove @ C ## Fs <=> notfunctor(C,rem),member(F,Fs),F==r | true.
```

The check notfunctor(C,rem) ensures that C is not a rem constraint. The check for set membership in the guards is expressed using the standard nondeterministic Prolog built-in predicate member/2.

Logical Retraction with killc/1. We extend the translation to allow for retraction of derived constraints. The constraint killc(C) logically retracts one occurrence of a constraint C. The two rules killc and killr try to find the constraint C. The killr rule applie sin the case where constraint C has been removed and is therefore now present in a rem constraint. The associated justifications point to all initial constraints that where involved in producing the constraint C. For retracting the constraint, it is sufficient to remove one of its producers. This introduces a choice which is implemented by the member predicate.

---

[2] More precisely, a simplification rule is generated if there are no kept constraints and a propagation rule is generated if there are no removed constraints.

```
killc @  killc(C), C ## Fs <=> member(F,Fs),F=r.
killr @  killc(C), rem(C ## FC) ## _Fs <=> member(F,FC),F=r.
```

Note that in the killr rule, we bind a justification F from FC, because FC contains the justifications of the producers of constraint C, while Fs also contains those that removed it by a rule application.

Dynamic Minimum Example. Translating the minimum rule to one with justifications results in:

```
min(A)##B \ min(C)##D <=> A<C | union([B,D],E), rem(min(C)##D)##E.
```

The following shows an example query and the resulting answer in SWI-Prolog:

```
?- min(1)##[A], min(0)##[B], min(2)##[C].
rem(min(1)##[A])##[A,B], rem(min(2)##[C])##[B,C], min(0)##[B].
```

The constraint min(0) remained. This means that 0 is the minimum. The constraints min(1) and min(2) have been removed and are now remembered. Both have been removed by the constraint with justification B, i.e. min(0).

We now logically retract with killc the constraint min(1) at the end of the query. The killr rule

```
killr @  killc(C), rem(C ## FC) ## _Fs <=> member(F,FC),F=r
```

applies and removes rem(min(1)##[A])##[A,B]. In the rule body, the justification A is bound to r – to no effect, since there are no other constraints with this justification:

```
?- min(1)##[A], min(0)##[B], min(2)##[C], killc(min(1)).
rem(min(2)##[C])##[B,C], min(0)##[B].
```

What happens if we retract the current minimum min(0)? The killc rule

```
killc @  killc(C), C ## Fs <=> member(F,Fs),F=r
```

applies, removes min(0)##[B] and binds justification B. The two rem constraints for min(1) and min(2) involve B as well, so these two constraints are re-introduced by applications of rule revive

```
revive @ rem(C##FC) ## Fs <=> member(F,Fs),F==r | C ## FC
```

The minimum rule applies to the two revived constraints. Note that min(2) is now removed by min(1) (before it was min(0)). The result is the updated minimum, which is 1:

```
?- min(1)##[A], min(0)##[B], min(2)##[C], killc(min(0)).
rem(min(2)##[C])##[A,C], min(1)##[B].
```

4   Optimizing the Implementation

We would like to avoid any overhead complexity-wise when computing with justifications as long as we do not use them for retraction. We are ready to accept a constant factor penalty. While the insertion of `rem` constraints takes constant time, the computation of the union of justifications is linear in the sizes of its input justification sets. The idea is to delay this computation until it is needed due to a retraction. We actually never compute the union of justifications, but will use the `union` constraints as data to find the necessary justifications. We describe the modifications for this new implementations and then discuss the complexity of this approach.

4.1   New Improved Implementation

To retract a constraint with justification `F`, the constraint `killd(F)` (kill down) finds its initial justifications. The arguments of the delayed `union` constraints are unbound variables now (except for the singleton sets of the justifications from the initial constraints in the query). The constraint `killd` has to find the `union` constraint with its justification in the output and follow all its input justifications (which are represented by a list). It proceeds recursively with the help of `kill1` (kill list) until it reaches an initial justification. On the way, we can stop if we see a justification again that we have already seen.

```
already_seen  @ killd(F) \ killd(F) <=> true.
go_to_initial @ union(FL,F) \ killd(F) <=> kill1(FL).

           kill1 @  kill1([]) <=> true.
           kill1 @  kill1([F|FL]) <=> killd(F), kill1(FL).
```

Then the auxiliary constraint `killone` (kill one) chooses one of these justifications in turn and removes it.

```
choice @ killone, killd([F]) <=> (F=r,waitrem ; killone).
done @   killone <=> false.
```

The rule `choice` uses Prolog's disjunction in the body. In the first disjunct, the binding of justification `F` to the constant `r` marks it as to be killed and wakes up all constraints in which this justification occurs. In this way, constraints are retracted and revived, respectively. The auxiliary constraint `waitrem` delays re-addition of previously removed constraints via the rule `revive` until all constraints with bound justification `F` have been retracted by the `remove` rule. This improves the performance. On Prolog's chronological backtracking, all computations caused by this first disjunct are undone and the second disjunct is tried. Note that `killd([F])` stays removed. The recursion on `killone` in the second disjunct ensures that all constraints `killd(F)` are eventually removed and thus all justifications are eventually tried. Note that as a consequence, in rule `done` for the base case when there are no more justifications, we must fail (not succeed), since we then have exhausted trying all justifications. Otherwise we would wrongly succeed with a computation that did not retract any constraint.

Now we also have to kill all output justifications of unions that have this killed justification as input justification, i.e. we go upwards.

```
go_upwards @  union(FL,F) <=> member(F1,FL),F1==[r] | F=[r].
```

Note that we will only pass a subset of the union constraints that killd visited, those that involve the chosen initial justification. We will also pass additional other union constraints as consequence of this.

Finally, for retraction, we remove constraints with killed justifications and we revive remembered constraints with killed justifications. We translate program constraints C with justifications F of the form c(X1,..Xn)##F into c(X1,..Xn,F) to support argument-wise indexing if necessary.

```
remove @ c(X1,..Xn,[r]) <=> true.
revive @ waitrem \ rem(c(X1,..Xn,FC),[r]) <=> c(X1,..Xn,FC).
        waitrem <=> true.
```

Here we put waitrem to work to trigger the re-addition of constraints in the revive rule. Having done so, waitrem is removed at the very end.

4.2   Worst-Case Time Complexity

We now discuss the complexity of our optimized implementation in terms of the input size and derivation length following the principles of [Frü02]. Let k be the largest number of head constraints in a given program. Note that k is a constant. Let c be the number of CHR constraints in the initial state (query). Let n be the derivation length of a computation, i.e. the number of rule applications (transitions).

The complexity of the original computation is at least n, because there are n rule applications that take at least constant time each. If the computation does not fail, each initial constraint is processed, which adds c to the lower bound of the complexity, which thus is $n + c$. Typically, n is larger than c, so we may assume just n.

All rule tries (application attempts) and rule applications take constant time, mostly because of the index on the justification. There is no overhead in runtime complexity until a constraint is killed: the union constraint and the rem constraints are just added to the constraint store. Since the number of rem constraints is bounded by k, complexity does not increase, if constraints can be added (inserted) in constant time. Based on these observations, we can also see that the space complexity is bounded by $O(n)$.

The union constraints have at most k input justifications that already have been introduced. The result is the output justification, represented by a new fresh logical variable. The union constraints form a directed acyclic graph (dag) with bounded width k, where the nodes are the justification set variables and where there is an directed edge (arc) from each input to the output justification for each union in a derivation. Since the output justification is always new, the corresponding graph is acyclic. It is typically not a tree, since a union may have input justifications from arbitrary previous unions.

There are at most n unions in a computation of length n. Thus there are at most n new justification nodes and c initial justifications. Therefore we have at most $n+c$ different nodes. The number of edges is at most k for each union and is therefore of order $O(n)$. The constraint killd has to go along at most kn edges, pass at most $n + c$ different nodes and stop at most at c initial justifications.

The constraint killone will chose the next initial justification in constant time. There may be up to c choices. Once we have chosen this initial justification to use for killing and retraction, we use the rule go_upwards to find all effected justifications with the help of the union constraint. We may have up to n non-initial justifications to revive and remove (kill) constraints in turn. Typically, the number will be much smaller, because n refers to all union constraints in the derivation. For each justification, there can only be a bounded number of remembered (k) and added constraints, because the number of head and body constraints in rules is bounded in a given program.

The killing of a justification and the retraction of constraints is accomplished by binding the justification variable. This will wake up all constraints in which the variable occurs. These are the union constraints and the all program constraints that have this justification. Thus the rule go_upwards and remove are immediately applicable, while the revive rule applications have to wait for the constraint waitrem.

In summary, the overall worst-case complexity of retracting a constraint with one choice of an initial justification is of order $O(n)$ (assuming $n > c$). The complexity trying each of the up to c found initial constraints is then $O(nc)$. Note that the complexity of removing all constraints or all initial c justifications in a computation is also bounded by $O(n)$, since the number of remembered and added constraints is also of order $O(n)$.

The additional cost of processing the revived re-added constraints is of course dependent on the given program and has to be added to the above complexity results. In the worst case, it amounts to a complete recomputation from scratch (cf. minimum example). It may be constant in the best case. If all rules of the program can be tried and applied in constant time, the derivation length n that was needed for c initial constraints may provide a $O(n)$ worst case complexity for computations with the revived constraints, thus leaving the overall worst-case complexity at $O(n)$.

5   Experiments

Experiments were run with SWI Prolog 6.2.1. in standard configuration on an Apple Mac Mini with OS X 10.9.5 2,5 GHz Intel Core i5 and 4 GB RAM. For compilation of the CHR files debugging was switched off and full optimization

enabled. We explicitly specified the arguments for indexing of program constraints in a declaration. This lead to a constant-factor improvement of the
runtime over automatic indexing provided by the CHR compiler.

We also introduced `passive` declarations in the rules that handle the justifications for retraction where feasible. These annotate head constraints in rules.
Such a constraint is then treated as data only that has to be searched for in
the constraint store. No active code is generated for that constraint, i.e. it does
not behave as an operation anymore that looks for its matching partner constraints. This optimizations avoids useless rule tries. Note that some of these
passive constraints are also automatically inferred by the compiler.

The programs used can be found in the appendix of the full online version
of this paper.

5.1 Dynamic All-Pair Shortest Paths

We want to find the shortest distance between all pairs of nodes in a complete
directed graph whose edges are annotated with non-negative distances. Initially,
for each edge, there is a corresponding path with the distance of the edge. For
every other pair of nodes, the unknown distances are initialized with ∞. Then
the following rule suffices to solve the problem:

```
shorten @ path(I,K,D1), path(K,J,D2) \ path(I,J,D3) <=>
                  D4 is D1+D2, D3>D4 | path(I,J,D4).
```

A currently shortest path between nodes I and J is replaced by the sum of the
distances between paths I to K and K to J if this new distance is shorter. Note
that the graph is complete. If the rule is not applicable anymore, all paths must
be shortest. From the `shorten` rule we generated the following rules augmented
with justifications

```
add_justification @ path(A,B,C) <=> path(A,B,C,[D]).
```

```
shorten @ path(A,B,C,D), path(B,E,F,G) \ path(A,E,H,I) <=>
     J is C+F,H>J |
         union([D,G,I],K), rem(path(A,E,H,I),K), path(A,E,J,K).
```

Example. The answer output has been slightly edited to improve readability.

```
?-path(a,b,1),path(b,a,2),path(a,c,3),path(c,a,0),path(b,c,1),path(c,b,4).
rem(path(c,b,4,[A]),B), rem(path(a,c,3,[C]),D), rem(path(b,a,2,[E]),F),
union([[G],[H],[A]],B), union([[H],[I],[C]],D), union([[I],[G],[E]],F),
path(c,b,1,B), path(a,c,2,D), path(b,a,1,F), path(b,c,1,[I]),
path(c,a,0,[G]), path(a,b,1,[H])
```

Initial justifications are in square brackets as single elements of lists. Thus the
last three paths in the answer were not shortened, while the other three paths

were shortened once, as can be seen by the deleted original path/3 constraints for them. From the first arguments of the delaying union/2 constraints we can also read off the constraints that lead to a shorter path.

For our experiments, the shorten rule was then instrumented to count rule tries (in the guard) and applications (in the body) with the help of Prolog's global variables. We explicitly added indexing information for the compiler because it slightly improved the performance on our examples. This means there is an hash index on the first and second argument of the path/4 constraint and it can also be accessed without index.

Random Graph Generation and Shortest Paths. We generated complete graphs from a given number of nodes represented by integers. For every pair of different nodes, a path is generated with a random distance between 1 and the number of nodes. This is accomplished by the rule:

```
gengraph(N), node(A), node(B) ==> random(1,N,D), path(A,B,D).
```

Previous Implementation				New Implementation		
Nodes	Apply	Try	Time	Nodes	Apply	Try	Time
12	125	2817	0.208	12	157	2958	0.106
12	113	2332	0.171	12	129	2770	0.093
12	142	2567	0.206	12	99	2362	0.083
14	210	4929	0.494	14	246	5215	0.225
14	250	5564	0.590	14	248	4693	0.189
14	223	4274	0.467	14	270	5449	0.234
16	338	9105	1.218	16	366	7667	0.402
16	379	8607	1.299	16	333	7643	0.391
16	362	8425	1.234	16	356	7613	0.391
18	501	11256	2.154	18	499	11899	0.759
18	502	12799	2.390	18	476	10567	0.674
18	416	9915	1.693	18	404	9980	0.628
21	801	21171	5.965	21	837	19134	1.499
21	783	22265	5.970	21	858	23550	1.928
21	778	19831	5.502	21	830	21094	1.676
24	1318	38188	14.809	24	1228	36507	3.553
24	1295	40549	16.172	24	1165	32543	3.422
24	1162	31898	12.270	24	1316	42426	4.039

Fig. 1. Shortest paths for random complete directed graphs

In Fig. 1 the number of nodes of the random directed graph is given, leading to a quadratic number of paths. Column *Apply* reports the number of applications of the shorten rule, while column *Try* shows how often this rule has been tried. Finally, *Time* reports the execution time in seconds. The time is roughly proportional to the number of rules tries indicating that indexing reduces the time for finding the three matching head constraints indeed to a constant.

Complexity. Let v be the number of nodes in the graph. There can be at most v^2 shortest path, one between each pair of nodes, so $c = v^2$. With indexes on the nodes in a path, the rule **shorten** can be applied in constant time, given one of the path constraints. The worst-case derivation length depends on the scheduling of paths for rule application. The optimal complexity is $O(v^3)$ when the scheduling of the Floyd-Warshall algorithm is used. It assumes an order on nodes and processes paths by their smallest nodes. We do not specify the scheduling and therefore expect a higher polynomial complexity in v. To reach the optimal complexity was not the scope of this work, since here we are interested in increasing the performance of logical retraction in comparison with the previous implementation.

Back to our experiments reported in Fig. 1: for a complete graph with v nodes and v^2 paths, the average execution time is of order $O(v^4)$ as was confirmed by computing the interpolating polynomial with WolframAlpha. This also holds for the number of rule tries and applications. So the derivation length n is quadratic in the number of paths c, i.e. $O(c^2)$. The previous implementation has a similar complexity, but a higher constant factor.

New Implementation						
Nodes	Apply	Try	Down	Up	Remove	Revive	Time
12	30	481	369	167	196	167	0.032
12	37	581	208	147	180	147	0.030
12	17	541	78	97	119	97	0.026
14	42	832	1990	242	281	242	0.075
14	54	894	1592	278	317	278	0.076
14	77	1245	1939	318	350	318	0.095
16	111	1901	5490	491	539	491	0.203
16	55	1413	3481	383	428	383	0.141
16	135	2513	1496	447	508	447	0.193
18	132	3158	7735	595	663	595	0.341
18	96	1869	4810	514	581	514	0.215
18	199	4598	8323	657	732	657	0.465
21	180	4963	51274	962	1033	962	1.170
21	207	4671	203119	917	966	917	2.980
21	188	4559	75441	908	985	908	1.381

Fig. 2. Removing all shortest paths from random graphs

Logical Retraction of Paths. In Fig. 2 we can see that the times for retracting all shortest paths in a complete random directed graph vary. The columns *Apply* and *Try* refer to accumulated recomputations of shortest paths after retraction of paths. *Down* reports the number of rule applications for going to the initial justifications through union constraints, while *Up* counts the propagation of the killed justification to the roots. The counts for *Down* and thus the time needed vary, the variation seems to increase the larger the graph is. This number depends on the number of updates to particular intermediate shortest paths, i.e on the depth of the justification dag.

Remove and *Revive* show the number of actual removals of constraints and re-addition of previously removed path constraints. These last two numbers are similar, with slightly more removals than revivals. (Note that re-added constraints may be removed afterwards.) The numbers for *Up* and *Revive* are identical, because the `shorten` rule always removes a single path constraint.

Overall, the complexity is once again quartic, $O(v^4)$. This corresponds once again to the derivation length and thus is in line with our complexity considerations in the previous section. It also means that the overhead of the recomputations is neglectable complexity-wise. Indeed, comparing the two figures, we can see that it typically takes less time to remove each shortest paths one by one and recompute all effected paths each time than to compute all the shortest paths initially. Moreover, the numbers of path recomputations are about a fourth of the number of initial path computations.

Note that recomputing from scratch would result in $O(v^2)$ recomputations (one for each retracted path) of complexity $O(v^4)$ each and thus in a polynomial of higher degree. The previous implementation also has a worse polynomial complexity for retracting constraints. For a graph of size 14, the previous implementation is already about an order of magnitude slower.

6   Conclusions

We presented an improved source-to-source transformation for logical retraction of constraints with justifications in CHR ($CHR^{\mathcal{J}}$). This transformation only imposes a constant factor overhead as long as justifications are not used for retraction. We argued that the worst-case time complexity for any number of retractions is in general proportional to the number of rule applications, i.e. derivation length. The complexity of an algorithm expressed in CHR is usually a polynomial in the derivation length. Therefore retraction indeed has typically less complexity than recomputation from scratch at the expense of storing removed constraints. The added space complexity is again bounded by the derivation length. In our experiments, we benchmarked the dynamic problem of maintaining shortest paths under addition and retraction of paths. The results support our complexity considerations. For future work, we would like to further improve the implementation and benchmark it. Since our rules for retraction are confluent with the original program [Fru17], we think that the optimization techniques in [AF04] might be helpful. At the same time, we would like to investigate how logical as well as classical algorithms like union-find behave when they become dynamic in $CHR^{\mathcal{J}}$.

References

[AF04] Abdennadher, S., Frühwirth, T.: Integration and optimization of rule-based constraint solvers. In: Bruynooghe, M. (ed.) LOPSTR 2003. LNCS, vol. 3018, pp. 198–213. Springer, Heidelberg (2004). https://doi.org/10.1007/978-3-540-25938-1_17

[DSdlBH04] Duck, G.J., Stuckey, P.J., de la Banda, M.G., Holzbaur, C.: The refined operational semantics of constraint handling rules. In: Demoen, B., Lifschitz, V. (eds.) ICLP 2004. LNCS, vol. 3132, pp. 90–104. Springer, Heidelberg (2004). https://doi.org/10.1007/978-3-540-27775-0_7

[Duc12] Duck, G.J.: SMCHR: satisfiability modulo constraint handling rules. Theory Pract. Log. Program. **12**(4–5), 601–618 (2012)

[FR18] Frühwirth, T., Raiser, F. (eds.): Constraint Handling Rules - Compilation, Execution, and Analysis. BOD, Norderstedt (2018). ISBN 9783746069050

[Frü02] Frühwirth, T.: As time goes by II: more automatic complexity analysis of concurrent rule programs. ENTCS **59**(3), 185–206 (2002). Di Pierro, A., Wiklicky, H. (eds.) QAPL 2001: Proceedings of First International Workshop on Quantitative Aspects of Programming Languages. Elsevier

[Frü09] Frühwirth, T.: Constraint Handling Rules. Cambridge University Press, Cambridge (2009)

[Frü15] Frühwirth, T.: Constraint handling rules - what else? In: Bassiliades, N., Gottlob, G., Sadri, F., Paschke, A., Roman, D. (eds.) RuleML 2015. LNCS, vol. 9202, pp. 13–34. Springer, Cham (2015). https://doi.org/10.1007/978-3-319-21542-6_2

[Fru17] Frühwirth, T.: Justifications in constraint handling rules for logical retraction in dynamic algorithms. In: Fioravanti, F., Gallagher, J.P. (eds.) LOPSTR 2017. LNCS, vol. 10855, pp. 147–163. Springer, Cham (2018). https://doi.org/10.1007/978-3-319-94460-9_9

[McA90] McAllester, D.A.: Truth maintenance. In: AAAI, vol. 90, pp. 1109–1116 (1990)

[SF06] Schrijvers, T., Frühwirth, T.: Optimal union-find in constraint handling rules, programming pearl. Theory Pract. Log. Program. (TPLP) **6**(1), 213–224 (2006)

[WGG00] Wolf, A., Gruenhagen, T., Geske, U.: On the incremental adaptation of chr derivations. Appl. Artif. Intell. **14**(4), 389–416 (2000)

[Wol01] Wolf, A.: Adaptive constraint handling with CHR in Java. In: Walsh, T. (ed.) CP 2001. LNCS, vol. 2239, pp. 256–270. Springer, Heidelberg (2001). https://doi.org/10.1007/3-540-45578-7_18

The Proportional Constraint
and Its Pruning

Armin Wolf[(⊠)]

IT4Energy Center, Fraunhofer FOKUS, Kaiserin-Augusta-Allee 31,
10589 Berlin, Germany
armin.wolf@fokus.fraunhofer.de

Abstract. Motivated by the necessity to model the energy loss of energy storage devices, a *Proportional Constraint* is introduced in finite integer domain Constraint Programming. Therefore rounding is used within its definition. For practical applications in finite domain Constraint Programming, pruning rules are presented and their correctness is proven. Further, it is shown by examples that the number of iterations necessary to reach a fixed-point while pruning depends on the considered constraint instances. However, fixed-point iteration always results in the strongest notion of bounds consistency. Furthermore, an alternative modeling of the Proportional Constraint is presented. The run-times of the implementations of both alternatives are compared showing that the implementation of the Proportional Constraint on the basis of the presented pruning rules performs always better on sample problem classes.

Keywords: Bounds consistency
Finite domain Constraint Programming · Fixed-point iteration
Proportional Constraint · Pruning rules

1 Motivation and Overview

Within the publicly funded project *WaveSave*[1] we are concerning cost-optimized trans-sectoral operation plans for hybrid energy systems within buildings. Those energy systems may consist of Combined Heat and Power (CHP) systems, Photovoltaic (PV) systems, heat pumps, boilers etc. as well as energy storage systems like hot water tanks or batteries. The operation of such systems is time critical and highly dynamic: Such systems have to react immediately to deviations in order to ensure the energy supply of the buildings and theirs users. Deviations might be caused by disturbances or uncertain forecasts. In order to generate cost-optimal operations plans (aka schedules) for the components of such hybrid energy systems in buildings, we model them as Constraint Optimization Problems (COP).

The presented work is funded by the German Federal Ministry for Economic Affairs and Energy within the project "WaveSave" (BMWi, funding number 03ET1312A).
[1] cf. http://www.it4energy-zentrum.de/de/it4energy/wavesave.

D. Seipel et al. (Eds.): DECLARE 2017, LNAI 10997, pp. 53–63, 2018.
https://doi.org/10.1007/978-3-030-00801-7_4

For an evaluation of configurations of energy systems in buildings with respect to their overall costs including investment, operation, emission, maintenance etc. over their live-time, we applied Mixed Integer Programming (MIP) while using similar approaches as presented in [1–3]. In our approach the MIP models are automatically generated from domain-specific XML descriptions defining the characteristics of the energy system components, the forecast data on energy use and supply, current states of the energy system components, etc. For MIP modeling the $<Coliop|Coin>$ *Mathematical Programming Language* (CMPL) (cf. http://www.coliop.org/) is used offering the opportunity to use different solvers like the open-source MIP solver CbC or the commercial MIP solver CPLEX.

To our knowledge it is rather difficult to consider domain-specific heuristics in MIP solvers, e.g. to consider preferences or soft constraints, e.g. in order to adapt schedules to changed constraints within the time-critical context of online operation scheduling of the energy systems components. Therefore, we decided to apply finite integer domain Constraint Programming (fdCP) allowing heuristic search to model and solve such COP adequately within this highly dynamic context. This means that good solutions have to be found or adapted within reasonable short time. Furthermore, Constraint Programming is the preferred choice in order to solve scheduling problems (cf. [4]).

Due to its nature, finite *integer* domain Constraint Programming only supports integer variables such that linear equations are considered as *diophantine equations*, i.e. only the integer solutions are sought. However, this is not suitable for any modeling of energy storage devices (cf., e.g. [2,3]) in the context of the *WaveSave* project, which requires the consideration of energy losses of energy storages to the environment over time.

Example 1. Let an energy storage be given having a characteristic energy loss of 3‰ within a given time unit. Further, let the energy load within time unit t be L_t of such a storage. Then the load L_{t+1} within the next time unit $t + 1$ is determined by at least the part which is proportional to the factor of loss, i.e. $L_{t+1} = 0.997 \cdot L_t + \cdots$.

Another situation in the energy context where the modeling of a proportional relationship with *diophantine equations* is not adequate occurs when the energetic behavior of CHP systems has to be modeled: Any CHP system has a specific *current characteristic* $\sigma > 0$ denoting the ratio of the supplied electric power P_{el} and the usable heat flow $\dot{Q}$.

Example 2. Let a Stirling engine based CHP system be given having a current characteristic $\sigma = 0.34$. Then it holds that the supplied electric power P_{el} is proportional to the usable heat flow $\dot{Q}$ which varies over time according to the operation mode of the CHP system:

$$P_{el} = 0.34 \cdot \dot{Q} \ .$$

Assuming that the energy loads of a storage or the electric and thermal powers of CHP systems over time are decision variables A and B, simple linear equations like $B = t \cdot A$ where $t \in \mathbb{R}$ are modeled in a straight-forward manner in any Linear Programming system. However, in fdCP this is not the case. There, the decision variables have integer domains such that only integer solutions are considered. Consequently, $L_{t+1} = 0.997 \cdot 45689$ has no integral solution. However $L_{t+1} = 45552 = \mathsf{round}(0.997 \cdot 45689)$ seems to be an acceptable approximation in this case.

The work is organized as follows: First we present some related work, then we define the *Proportional Constraint* and some pruning rules. Further, the correctness of these rules is proven and it is shown that iterative pruning leads to the strongest notion of bounds consistency. Then, an alternative modeling of the *Proportional Constraint* based on linear inequalities is presented and the run-times of these two approaches on sample problem classes are compared. Finally we conclude with some remarks on the implementation and the use of this constraint.

2 Related Work

Linear equations $y = \alpha \cdot x$ where the variable y *is proportional to* another variable x – where α is a scalar value – are special cases of *weighted sums*, i.e. $y = \alpha_1 \cdot x_1 + \cdots + \alpha_n \cdot x_n (n > 0)$. *Weighted sum constraints* are already considered in [5]. Applying the pruning rules defined there on finite domain integer variables, the resulting consistency only ensures that there exist *real* solutions which is not adequate in the context of our *WaveSave* project. There, the solutions have to integral, i.e. the integer values of finite domain decision variables considered in further constraints. Thus, we decided to extend our object-oriented constraint solving library firstCS [6] which already supports *weighted sum constraints* with an adequate *proportional constraint* $y = \alpha \cdot x$ for *finite domain integer variables*.

3 The Proportional Constraint

In application domains of finite domain Constraint Programming such as the optimized operation of energy systems there is a need to model a *proportional* energy loss when using energy storage devices as already mentioned. Therefore and for other applications as well, we define the binary *Proportional Constraint*:

Definition 1 (Proportional Constraint). *Let $t > 0$ be a real value and A, B finite domain constraint variables having integer domains* $\mathsf{dom}(A)$ *respective* $\mathsf{dom}(B)$. *For convenience, let* $\min(X) = \min(\mathsf{dom}(X))$ *and* $\max(X) = \max(\mathsf{dom}(X))$ *for any domain variable X. The (binary) Proportional Constraint*

$$\mathsf{round}(t \cdot A) = B$$

is satisfied, if for any value $a \in \mathsf{dom}(A)$ there is a value $b \in \mathsf{dom}(B)$ respective if for any value $b \in \mathsf{dom}(B)$ there is a value $a \in \mathsf{dom}(A)$ such that $\mathsf{round}(t \cdot a) = b$

holds. Such value pairs (a, b) *or* labelings $\Theta = \{A \mapsto a, B \mapsto b\}$ *satisfying the constraint are called* solutions. *There,* round(.) *is the* rounding *function as defined by* round$(x) = \lfloor x + 0.5 \rfloor$, *where* $\lfloor y \rfloor$ *is the greatest integer value less than or equal to* y *for any real value* y.

The definition of the *Proportional Constraint* is *sound* in the sense that for any $t > 0$ and any integer value a there is another integer value b such that round$(t \cdot a) = b$ holds: For a there is obviously $b = $ round$(t \cdot a)$. For any integer value b and $0 < t \leq 1$ there is always an integer value $a \in [(b-0.5)/t, (b+0.5)/t]$ satisfying the constraint, because it holds

$$t \leq 1$$
$$\Leftrightarrow \quad t \leq (b + 0.5) - (b - 0.5)$$
$$\Leftrightarrow \quad 1 \leq \frac{b + 0.5}{t} - \frac{b - 0.5}{t}$$

However, for $t > 1$ this is not always the case: Let $b = 1$ and $t = 1.9$, then there is not any integer value a such that $1 = $ round$(1.9 \cdot a)$ holds.

In order to implement and use such a constraint in a Constraint Programming system some *pruning rules* have to be defined, reducing the domains of the involved variables without losing any solutions and resulting in a fixed-point when iterated such that the pruned domains of the variables hopefully satisfy some notion of consistency. Our definition of some pruning rules for the Proportional Constraint requires a special kind of "floor" function mapping reals to integers. It is defined as follows:

Definition 2. *For any real value* x *let the function* $\lfloor . \rfloor : \mathbb{R} \to \mathbb{Z}$ *be defined by*

$$\lfloor x \rfloor = \begin{cases} x - 1 & \text{if } x = \lfloor x \rfloor, \\ \lfloor x \rfloor & \text{otherwise.} \end{cases}$$

This definition of the function $\lfloor . \rfloor$ is sound in the sense that for any $x \in \mathbb{R}$ there is exactly one $y \in \mathbb{Z}$ such that $y = \lfloor x \rfloor$ holds.

For the defined Proportional Constraint we propose the following *pruning rules*:

Definition 3 (Pruning Rules). *For any Proportional constraint*

$$\text{round}(t \cdot A) = B$$

with $t > 0$ *and finite domain constraint variables* A *and* B *having integer domains* dom(A) *respective* dom(B) *let*

$$\text{dom}^*(B) = \text{dom}(B) \cap [\text{round}(t \cdot \min(A)), \text{round}(t \cdot \max(A))] \qquad (1)$$
$$\text{dom}^*(A) = \text{dom}(A) \cap [\lceil (\min^*(B) - 0.5)/t \rceil, \lfloor (\max^*(B) + 0.5)/t \rfloor] \qquad (2)$$

be *some* pruning rules – *to be applied in the given order* – *where* $\min^*(B) = \min(\text{dom}^*(B))$ *and* $\max^*(B) = \max(\text{dom}^*(B))$ *for the sake of convenience. Further, let* $\lceil y \rceil$ *be the smallest integer value greater than or equal to y for any real value y.*

These rules are potentially reducing the domains of A and B, i.e. $\text{dom}^*(A) \subseteq \text{dom}(A)$ *and* $\text{dom}^*(B) \subseteq \text{dom}(B)$ *hold.*

The indicator $*$ *in* $\text{dom}^*(A)$ *respective in* $\text{dom}^*(B)$ *is used to distinguish between the original and the updated domains of the variables A and B which will replace* $\text{dom}(A)$ *respective* $\text{dom}(B)$ *in any next iteration of these pruning rules.*

Example 3. Let the Proportional Constraint $\text{round}(2.1 \cdot A) = B$ with $\text{dom}(A) = \{0, 1, 2, 3\}$ and $\text{dom}(B) = \{2\}$ be given. After applying the pruning rules defined in Definition 3 it holds that $\text{dom}^*(B) = \{2\} \cap [0, 6] = \{2\}$ and $\text{dom}^*(A) = \{0, 1, 2, 3\} \cap [1, 1] = \{1\}$. This means that pruning determines the solution $\Theta = \{A \mapsto 1, B \mapsto 2\}$ correctly.

Obviously, the question arises whether the pruning rules are in general *correct* or whether there are any integer solutions of the constraint which will be lost while pruning? – The following proposition answers this question:

Proposition 1. *Let* $t > 0$ *be a real value and* A, B *finite domain constraint variables having integer domains* $\text{dom}(A), \text{dom}(B)$. *If there is a value pair* $(a, b) \in \text{dom}(A) \times \text{dom}(B)$ *such that* $b = \text{round}(t \cdot a)$ *holds, then it will hold that* $(a, b) \in \text{dom}^*(A) \times \text{dom}^*(B)$. *In other words the propagation rules are correct, i.e. not any integer solution is lost while pruning.*

Proof. Let $a \in \text{dom}(A)$ and $b \in \text{dom}(B)$ be any two integer values such that $\text{round}(t \cdot a) = b$ holds. Due to the fact that $\min(A) \le a \le \max(A)$ holds and the function $f(x) = \text{round}(t \cdot x))$ is monotonic, it holds $f(\min(A)) \le f(a) \le f(\max(A))$ and thus $b \in \text{dom}^*(B)$. According to the definition of round exactly one of the following two cases is valid:

1. $t \cdot a = \text{round}(t \cdot a) + \varepsilon$ with $0 \le \varepsilon < 0.5$,
2. $t \cdot a = \text{round}(t \cdot a) - \varepsilon$ with $0 < \varepsilon \le 0.5$.

Let us suppose that the first case is valid. Due to the facts that $b \in \text{dom}^*(B)$ and $b = \text{round}(t \cdot a)$ it holds that

$$
\begin{aligned}
\lceil (\min^*(B) - 0.5)/t \rceil &\le \lceil (b - 0.5)/t \rceil \\
&= \lceil (\text{round}(t \cdot a) - 0.5)/t \rceil \\
&= \lceil ((t \cdot a - \varepsilon) - 0.5)/t \rceil \\
&= \lceil a - (0.5 + \varepsilon)/t \rceil \\
&\le a \ .
\end{aligned}
\tag{3}
$$

Further, it holds that

$$\lfloor(\max{}^*(B) + 0.5)/t\rfloor \geq \lfloor(b + 0.5)/t\rfloor$$
$$= \lfloor(\mathrm{round}(t \cdot a) + 0.5)/t\rfloor$$
$$= \lfloor(t \cdot a - \varepsilon) + 0.5)/t\rfloor$$
$$= \lfloor a + (0.5 - \varepsilon)/t\rfloor$$
$$\geq a \tag{4}$$

given that $(0.5 - \varepsilon)/t > 0$ holds.

Now, let us suppose that the second case is valid. Due to the facts that $b \in \mathrm{dom}^*(B)$ and $b = \mathrm{round}(t \cdot a)$ it holds that

$$\lceil(\min{}^*(B) - 0.5)/t\rceil \leq \lceil(b - 0.5)/t\rceil$$
$$= \lceil(\mathrm{round}(t \cdot a) - 0.5)/t\rceil$$
$$= \lceil(t \cdot a + \varepsilon) - 0.5)/t\rceil$$
$$= \lceil a - (0.5 - \varepsilon)/t\rceil$$
$$\leq a \tag{5}$$

given that $(0.5 - \varepsilon)/t \geq 0$ holds. Further it holds that

$$\lfloor(\max{}^*(B) + 0.5)/t\rfloor \geq \lfloor(b + 0.5)/t\rfloor$$
$$= \lfloor(\mathrm{round}(t \cdot a) + 0.5)/t\rfloor$$
$$= \lfloor(t \cdot a + \varepsilon) + 0.5)/t\rfloor$$
$$= \lfloor a + (0.5 + \varepsilon)/t\rfloor$$
$$\geq a \; . \tag{6}$$

$\square$

An iterated application of the pruning rules defined in Definition 3 on the constraint variables' domains either reduces these finite domains until they become empty or will not be further reduced. In any case, the iteration stops after a finite number of steps, such that $\mathrm{dom}^*(A) = \mathrm{dom}(A)$ and $\mathrm{dom}^*(B) = \mathrm{dom}(B)$ holds, i.e. any further applications of the pruning rules will not change the domains of the variables. This means that a *fixed-point* is reached (cf. [7]). However, how many iterations are necessary for reaching a fixed-point? – The following proposition answers this question:

Proposition 2. *The number of iterations of the pruning rules necessary to reach a fixed-point has no fixed upper bound: The number of iterations strongly depends on the constraint instance, in particular on the structure and on the size of the domains of the variables.*

Example 4 (Counter Examples). Let $t = 3.0$ and for any integer value $n > 1$ let $\mathrm{dom}(A) = \{1, 2, 3, \ldots, n\}$ and $\mathrm{dom}(B) = \{3, 5, 8, \ldots, 3n - 1\}$ be given. Then a fixed-point is reached after at least $n - 1$ iterations. The same holds for $t = 0.3$

and any integer value $n > 1$ if $\mathsf{dom}(A) = \{10, 20, 30, \ldots, 10 \cdot n\}$ and $\mathsf{dom}(B) = \{3, 5, 8, \ldots, 3n - 1\}$, then a fixed-point is reached after $n - 1$ iterations, too.[2]

Finally we show that after a fixed-point iteration of the pruning rules (cf. Definition 3) the domains of the variables of the Proportional Constraint are *bounds consistent* in the strongest sense – cf. [8] for a detailed analysis of different notions of *bounds consistency*. From there we adopted the following definition:

Definition 4. *A domain D is bounds($\mathcal{D}$) consistent for a constraint c where $\mathsf{vars}(c) = \{x_1, \ldots, x_n\}$, if for each variable x_i with $1 \leq i \leq n$ and for each $d_i \in \{min(x_i), max(x_i)\}$ there exist integers d_j with $d_j \in \mathsf{dom}(x_j)$ where $1 \leq j \leq n$, $j \neq i$ such that the labeling $\Theta = \{x_1 \mapsto d_1, \ldots, x_n \mapsto d_n\}$ is an integer solution of c.*

This definition considers n-ary constraints and thus binary constraints like the Proportional Constraint as well.

Proposition 3. *Let a Proportional Constraint*

$$\mathsf{round}(t \cdot A) = B$$

be given with $t > 0$ and finite domain constraint variables A and B having integer domains $\mathsf{dom}(A)$ respective $\mathsf{dom}(B)$. Furthermore, it is assumed that the pruning rules (cf. Definition 3) are iterated until a fixed-point is reached, i.e. it holds that $\mathsf{dom}^*(A) = \mathsf{dom}(A)$ and $\mathsf{dom}^*(B) = \mathsf{dom}(B)$. Then it holds that the domain D (cf. Definition 4) consisting of $\mathsf{dom}(A)$ and $\mathsf{dom}(B)$ is bounds($\mathcal{D}$) consistent.

Proof. Let $a = \min(A)$. Then, there is an integer value b such that $b = \mathsf{round}(t \cdot a)$ holds. Now we assume that $b \notin \mathsf{dom}(B)$ respective that $b \neq \min(B)$. It follows that $b < \min(B)$ due to the pruning rule (1) and the monotonicity of the rounding function. Thus, $\mathsf{round}(t \cdot a) \leq \min(B) - 1$ holds and further $t \cdot a \pm \varepsilon \leq \min(B) - 1$ (cf. case distinction in the proof of Proposition 1). This implies that $a < (\min(B) - 0.5)/t$ and finally $a < \lceil (\min(B) - 0.5)/t \rceil$ holds. This contradicts $a = \min(A)$, i.e. the assumption is wrong, it holds that $b \in \mathsf{dom}(B)$ respective that $b = \min(B)$.

Let $a = \max(A)$. Then, there is an integer value b such that $b = \mathsf{round}(t \cdot a)$ holds. Now we assume that $b \notin \mathsf{dom}(B)$ respective that $b \neq \max(B)$. It follows that $b > \max(B)$ due to the pruning rule (1) and the monotonicity of the rounding function. Thus, $\mathsf{round}(t \cdot a) \geq \max(B) + 1$ holds and further $t \cdot a \pm \varepsilon \geq \max(B) + 1$ (cf. case distinction in the proof of Proposition 1). This implies that $a \geq (\max(B) + 0.5)/t$ and finally $a > \lfloor (\max(B) + 0.5)/t \rfloor$ because a is an integer value. This contradicts $a = \max(A)$, i.e. the assumption is wrong, it holds that $b \in \mathsf{dom}(B)$ respective that $b = \max(B)$.

Let $b = \min(B)$. We further distinguish two additional sub-cases:

---

[2] The formal proof by induction is left to the interested reader.

(a) We further suppose that there is an integer value a such that $b = \text{round}(t \cdot a)$ holds. Now, we assume that $a \notin \text{dom}(A)$ respective that $a \neq \min(A)$. It follows that $a < \min(A)$ due to the pruning rule (2) and the monotonicity of the rounding function. Thus, $t \cdot a < \min(B) - 0.5$ holds implying that $b = \text{round}(t \cdot a) \leq t \cdot a + 0.5 < \min(B)$, i.e. the assumption was wrong, it holds that $a \in \text{dom}(A)$ respective that $a = \min(A)$.

(b) Now, we assume that for each integer value a it holds that $b \neq \text{round}(t \cdot a)$ even for $a = \min(A)$. Thus, $b > \text{round}(t \cdot \min(A))$ holds. According to the case distinction in the proof of Proposition 1 it holds that $b \geq t \cdot \min(A) \pm \varepsilon + 1$ and thus in either case $b \geq t \cdot \min(A) + 0.5$ holds. It follows that $(\min(B) - 0.5)/t > \min(A)$ and thus $\lceil (\min(B) - 0.5)/t \rceil > \min(A)$ are holding. This contradicts the pruning rule (2). The assumption is wrong, i.e. there is an integer value a such that $b = \text{round}(t \cdot a)$ holds. The case (b) never occurs.

Let $b = \max(B)$. Again, we further distinguish two additional sub-cases:

(a) We further suppose that there is an integer value a such that $b = \text{round}(t \cdot a)$ holds. Now, we assume that $a \notin \text{dom}(A)$ respective that $a \neq \max(A)$. It follows that $a > \max(A)$ due to the pruning rule (2) and the monotonicity of the rounding function. Thus, $t \cdot a \leq \max(B) + 0.5$ holds. Due to the fact that $\max(B)$ is an integer value it holds that $b = \text{round}(t \cdot a) > \max(B)$, i.e. the assumption was wrong, it holds that $a \in \text{dom}(A)$ respective that $a = \max(A)$.

(b) Now, we assume that for each integer value a it holds that $b \neq \text{round}(t \cdot a)$ even for $a = \max(A)$. Thus, $b < \text{round}(t \cdot \max(A))$ holds. According to the case distinction in the proof of Proposition 1 it holds that $b \leq t \cdot \max(A) \pm \varepsilon - 1$ and thus in either case $b \leq t \cdot \max(A) - 0.5$. It follows that $(\max(B) + 0.5)/t \leq \max(A)$ and thus $\lfloor (\max(B) + 0.5)/t \rfloor < \max(A)$ are holding because $\max(A)$ is an integer value. This contradicts the pruning rule (2). The assumption is wrong, i.e. there is an integer value a such that $b = \text{round}(t \cdot a)$ holds. The case (b) never occurs.

$\square$

4 Alternative Modeling of the Proportional Constraint

For any *rational* factor $t > 0$ within a Proportional Constraint $\text{round}(t \cdot A) = B$ it is possible to model this constraint equivalently on the basis of a *weighted sum* constraints which are well established in fdCP [5]:[3]

Proposition 4. *Let $t = p/q$ where p and q are positive integer values. Then, for any two finite domain constraint variables A, B the constraint*

$$-\frac{q}{2} < q \cdot B - p \cdot A \leq \frac{q}{2} \tag{7}$$

---

[3] Many thanks to the anonymous reviewer who suggested this approach.

is equivalent to the Proportional Constraint $\mathsf{round}(t \cdot A) = B$ *in the sense that any solution* $\{A \mapsto a, B \mapsto b\}$ *is a solution of (7) and vice-versa.*

Proof. Let $\{A \mapsto a, B \mapsto b\}$ be any solution of the Proportional Constraint, i.e. $b = \mathsf{round}(t \cdot a)$. We distinguish two sub-cases:

1. Let $q = 1$, i.e. t be an integer value. It holds that $q{\cdot}b - p{\cdot}a = \mathsf{round}(p{\cdot}a) - p{\cdot}a = 0$. Obviously, $\{A \mapsto a, B \mapsto b\}$ is a solution of (7).
2. Let $q > 1$, i.e. t be non-integral. According to the case distinction in the proof of Proposition 1 it holds that $q \cdot b - p \cdot a = q \cdot \mathsf{round}(p/q \cdot a) - p \cdot a = q \cdot p/q \cdot a \pm \varepsilon - p \cdot a = \pm\varepsilon$. Due to the fact that $-q/2 < -\varepsilon$ and $\varepsilon \le q/2$ holds for any $q > 1$, it also holds that $\{A \mapsto a, B \mapsto b\}$ is a solution of (7).

Now, let $\{A \mapsto a, B \mapsto b\}$ be any solution of (7), i.e. $-q/2 < q{\cdot}b - p{\cdot}a \le q/2$ holds and thus $-1/2 < b - p/q \cdot a \le 1/2$. Consequently, $b = \mathsf{round}(t \cdot a)$ is the only integer value satisfying this condition and thus $\{A \mapsto a, B \mapsto b\}$ is also a solution of the Proportional constraint. □

5   Run-Time Comparison

For a run-time comparison of the Proportional Constraint and its alternative modeling (cf. (7)) we implemented the Proportional Constraint with the pruning rules presented in Definition 3 in our finite domain constraint solving library firstCS [6] which already supports *weighted sum constraints*. Then, we modeled the following classes of problem instances

- A(n): $t = 0.3$, $\mathsf{dom}(A) = \{10, 20, 30, \ldots, 10n\}$, $\mathsf{dom}(B) = \{3, 5, 8, \ldots, 3n-1\}$.
- B(n): $t = 0.997$, $\mathsf{dom}(A) = \{1, 2, 3, \ldots, n\}$, $\mathsf{dom}(B) = \{1, 2, 3, \ldots, n\}$.
- C(n): $t = 0.003$, $\mathsf{dom}(A) = \{1, 2, 3, \ldots, n\}$, $\mathsf{dom}(B) = \{1, 2, 3, \ldots, n\}$.

For problem class A (cf. Example 4) we perform initial pruning resulting in one solution. For the problem classes B and C we perform initial pruning and additional pruning while searching all solutions. The search strategy chooses variable A then B and their values in increasing order performing chronological backtracking in cases where dead ends are reached, i.e. a domain of a variable becomes empty. Pruning means that the according pruning rules are applied until a fixed-point is reached.

Table 1 shows the results of our run-time comparison of the Proportional Constraint (PC) and its alternative modeling (ALT) executed on a Windows computer with Windows 10 Pro (64 bit), Intel i7 CPU, 2.60 GHz, 12 GByte RAM. The computations for all problem instances were repeated 10 times. We compared best run-times and average run-times (in ms) showing that the execution of the Proportional Constraint is always faster – in the best cases 79%, in the worst case 2% and on average 45%:

Table 1. Run-time comparison of the Proportional Constraint and its alternative

Instance	PC avg.	PC best	ALT avg.	ALT best	ALT/PC avg.	ALT/PC best
A(10000)	310.2	297	539.5	440	174%	148%
A(20000)	1238.8	1199	1277.8	1227	**103%**	**102%**
A(40000)	4035.3	3999	4985.4	4864	124%	122%
A(80000)	15582.6	15280	19188.7	18885	123%	124%
B(10000)	103.3	78	170.7	109	165%	140%
B(20000)	149.5	125	200.1	172	134%	138%
B(40000)	196.3	187	278.3	250	142%	137%
B(80000)	276.5	250	452.9	406	164%	162%
C(10000)	86.0	78	153.6	140	**179%**	**179%**
C(20000)	123.4	109	218.8	156	177%	143%
C(40000)	191.4	172	281.5	265	147%	154%
C(80000)	273.4	250	409.7	359	150%	144%

6   Conclusion

Within this work a *Proportional Constraint* for finite integer domains is defined and according pruning rules are presented and analyzed. Its is shown that pruning based on these rules is correct and results in the strongest notion of bounds consistency (cf. [8]). The introduced *Proportional Constraint* is implemented in our object-oriented constraint solving library firstCS [6] and compared with an alternative approach based on linear inequalities already available in firstCS.

The Proportional Constraint is used in the context of the *WaveSave* project to model the energy loss of energy storages like heat tanks or batteries over time and the relationship between the supplied electric powers and the usable heat flows of CHP systems. It is noteworthy that for other applications we implemented a more general version of the *Proportional Constraint* for any $t \in \mathbb{R}$: For $t = 0$ the pruning rules are trivial: If $0 \notin \text{dom}(B)$ holds, $\text{dom}^*(B)$ will become empty as well as $\text{dom}^*(A)$. If $0 \in \text{dom}(B)$ holds, $\text{dom}^*(B) = \{0\}$ and $\text{dom}^*(A) = \text{dom}(A)$ will hold. For $t < 0$ the pruning rules for $t > 0$ are adapted accordingly respecting the fact that $B = \text{round}(-t \cdot -A)$ while pruning the domain of A and $-B = \text{round}(-t \cdot A)$ while pruning the domain of B.

References

1. Bosman, M., Bakker, V., Molderink, A., Hurink, J., Smit, G.: Planning the production of a fleet of domestic combined heat and power generators. Eur. J. Oper. Res. **216**, 140–151 (2012)
2. Bozchalui, M.C., Sharma, R.: Optimal operation of commercial building microgrids using multi-objective optimization to achieve emissions and efficiency targets. In: 2012 IEEE Power and Energy Society General Meeting, pp. 1–8. IEEE (2012)

3. Brahman, F., Honarmand, M., Jadid, S.: Optimal electrical and thermal energy management of a residential energy hub, integrating demand response and energy storage system. Energy Build. **90**, 65–75 (2015)
4. Baptiste, P., Pape, C.L., Nuijten, W.: Constraint-Based Scheduling - Applying Constraint Programming to Scheduling Problems. Springer, Boston (2001). https://doi.org/10.1007/978-1-4615-1479-4
5. Schulte, C., Stuckey, P.J.: When do bounds and domain propagation lead to the same search space? ACM Trans. Program. Lang. Syst. (TOPLAS) **27**(3), 388–425 (2005)
6. Wolf, A.: firstCS - new aspects on combining constraint programming with object-orientation in Java. KI - Künstliche Intelligenz **26**(1), 55–60 (2012)
7. Apt, K.R.: From chaotic iteration to constraint propagation. In: Degano, P., Gorrieri, R., Marchetti-Spaccamela, A. (eds.) ICALP 1997. LNCS, vol. 1256, pp. 36–55. Springer, Heidelberg (1997). https://doi.org/10.1007/3-540-63165-8_163
8. Choi, C.W., Harvey, W., Lee, J.H.M., Stuckey, P.J.: Finite domain bounds consistency revisited. In: Sattar, A., Kang, B. (eds.) AI 2006. LNCS (LNAI), vol. 4304, pp. 49–58. Springer, Heidelberg (2006). https://doi.org/10.1007/11941439_9

An Operational Semantics
for Constraint-Logic Imperative
Programming

Jan C. Dageförde(✉) and Herbert Kuchen

ERCIS, University of Münster, Münster, Germany
{dagefoerde,kuchen}@uni-muenster.de

Abstract. Object-oriented (OO) languages such as Java are the dominating programming languages nowadays, among other reasons due to their ability to encapsulate data and operations working on them, as well as due to their support of inheritance. However, in contrast to constraint-logic languages, they are not particularly suited for solving search problems. During development of enterprise software, which occasionally requires some search, one option is to produce components in different languages and let them communicate. However, this can be clumsy.

As a remedy, we have developed the constraint-logic OO language Muli, which augments Java with logic variables and encapsulated search. Its implementation is based on a symbolic Java virtual machine that supports constraint solving and backtracking. In the present paper, we focus on the non-deterministic operational semantics of an imperative core language.

Keywords: Java · Operational semantics · Encapsulated search
Programming paradigm integration

1 Introduction

Contemporary software development is dominated by object-oriented (OO) programming. Its programming style benefits most industry applications by providing e.g. inheritance and encapsulation of structure and behaviour, since these concepts can positively contribute towards reusability and maintainability [13]. Nevertheless, some industry applications require search, for which constraint-logic programming is more suited than OO (or imperative) programming. However, developing applications that integrate both worlds, e.g. a Java application using a Prolog search component via Java Native Interface (JNI), is tedious and error-prone [10].

For that reason, we propose the *Münster Logic-Imperative Programming Language (Muli)*, integrating constraint-logic programming with OO programming in a novel way. Based on Java, it adds logic variables and encapsulated search to

© Springer Nature Switzerland AG 2018
D. Seipel et al. (Eds.): DECLARE 2017, LNAI 10997, pp. 64–80, 2018.
https://doi.org/10.1007/978-3-030-00801-7_5

the language, supported by constraint solvers and non-deterministic execution on a symbolic Java virtual machine (JVM). The symbolic JVM adapts concepts from the Warren Abstract Machine, such as choice points and trail [22]. Muli's tight integration of both paradigms facilitates development of applications whose business logic is implemented in Java, but which also require occasional search, such as operations research applications [8].

In this paper, we describe a reduction semantics for a core subset of Muli. In particular, the interaction of imperative statements, free variables, and non-determinism is of interest. For simplicity, this core language abstracts from inheritance, multi-threading, and reflection, because those features do not exhibit interesting behaviour w.r.t. our semantics. The formulated semantics is helpful to get an understanding of the mechanics behind concepts that are novel to imperative and OO programming, and serves as a formal basis for implementing the symbolic JVM. It can also be used for reasoning about applications developed in Muli.

To that end, our paper is structured as follows. We provide an overview of the new language and its concepts in Sect. 2. Section 3 formalises the operational semantics of the core language. An example evaluation using this semantics is shown in Sect. 4. Section 5 presents a discussion of our concepts. Related work is outlined in Sect. 6. We then conclude in Sect. 7 and provide an outlook towards further research.

2   Language Concepts

The Muli language is derived from Java 8. We do not change existing concepts and features of Java, so that Muli also benefits from Java's well-known and well-received features, such as OO and managed memory. Instead, the language is defined by its additions to Java, i.e. Muli is a superset of Java.

Muli adds the concept of *free variables*, i.e. variables that are declared and instantiated, but not to a particular value. Instead, they are treated symbolically and can be used in statements and expressions. *Constraints* on symbolic variables and expressions are imposed during symbolic execution of conditional statements. For example, an if statement with a condition that involves insufficiently constrained variables results in multiple branches that can be evaluated. Conceptually, we can non-deterministically choose a branch and evaluate it. Our implementation considers all these branches using backtracking and a (complete!) iterative deepening depth-first search strategy. This is supported by a specialised symbolic JVM that records choice points for each non-deterministic branch.

Furthermore, we enforce that non-determinism only takes place inside *encapsulated search* regions, whereas code outside encapsulation is executed deterministically. This ensures that non-determinism is not introduced by accident, intending not to harm the understanding of known Java concepts. Furthermore, this ensures that the overall application exits in a single state. In contrast, unencapsulated symbolic execution could result in multiple exit states, which could cause difficulties on the side of the caller. Encapsulation is expressed by using either of the getAllSolutions and getOneSolution operators. The logic of encapsulated search is described by *search regions* that are implemented as methods, e.g. as lambda abstractions, in order to defer their evaluation until encapsulation begins.

Solutions of encapsulated search are defined by values or expressions returned from search regions. Due to non-determinism, multiple solutions can be returned from search. Additionally, we introduce the special statement **fail**;, whose evaluation results in immediate backtracking in the symbolic JVM without recording a solution for the current branch.

From a syntactic perspective, these concepts extend Java only minimally. The resulting syntax of Muli can best be demonstrated using an example. Listing 1 exhibits a Muli method log() that searches for the logarithm of a number x to the base 2 using a free variable y and a method pow that calculates b^y imperatively, which is constrained to be equal to x.

```
int log(int x) {
    int y free;
    if (pow(2,y) == x) return y;
    else fail; }
int pow(int b, int y) {
    int i; int r; i = 0; r = 1;
    while (i < y) {
        r = r * b; i = i + 1; }
    return r; }
```

Listing 1. Non-deterministic computation of the logarithm of a number to the base 2 using (core) Muli.

Let us assume that the considered search region consists of a call to log, e.g. log(4). When calling log with a given x, the free variable y is created and then passed to pow that calculates the power b^y symbolically, as y is free. Therefore, it returns a value that is accompanied by a set of accumulated constraints from which this particular value follows.[1] Consequently, log computes the logarithm by defining a constraint system using an imperative method that calculates the power.

If the variables involved in a branching condition (of **if** or **while** in Listing 1) are not sufficiently constrained, one of the feasible branches is chosen non-deterministically. Actually, our symbolic JVM would try them systematically one after the other, aided by a backtracking mechanism. When selecting a branch, the corresponding condition is added to the constraint store and consistency is checked. For example, **while** (i < y) can be either true or false as y is a free variable. As a result, one branch assumes the condition to be true and therefore adds the constraint $i < y$ to the constraint store by imposing a conjunction of the existing store and the new constraint. In contrast, the second branch

---

[1] In other problems the return value could be a symbolic expressions if the accumulated constraints do not reduce the return value's domain to a concrete value.

assumes it to be false and therefore adds the negated condition as a constraint. If an added constraint renders the store inconsistent, backtracking occurs, i.e. that branch is pruned and execution continues with a subsequent branch. Similarly, backtracking occurs when a solution is found so that the next branch can be evaluated to find further solutions. Muli's encapsulated search operators use lazy streams to return collected solutions to the surrounding deterministic computation, such that the surrounding computation can decide how many solutions it wants to obtain.

3   A Non-deterministic Operational Semantics of Muli

Muli is an extension to Java and therefore intends to fully support all Java functionality. In fact, all Muli programs even compile to regular JVM bytecode that can be parsed and executed by a regular JVM (but incorrectly), and all Java programs can be executed correctly by Muli's symbolic JVM. Outside of encapsulated search, execution in Muli is deterministic and replicates the behaviour of a standard JVM [12]. Inside encapsulation, search regions are executed non-deterministically. This changes the semantics of Java and adds subtleties that need to be explicated, particularly regarding the interaction of imperative statements, free variables, and non-determinism. Therefore, we formally define the semantics for non-deterministic evaluation of search regions.

For the purpose of describing a (non-deterministic) operational semantics of Muli, we focus on an imperative, procedural subset of Java (and Muli). This concise subset allows us to focus on the interaction between imperative and constraint-logic programming. It therefore abstracts from some features that are expected from Java but that would not contribute to the discussion in the present paper, such as inheritance.[2] Furthermore, this semantics abstracts from the execution of deterministic program parts and therefore does not prescribe an implementation for the encapsulation operators, `getAllSolutions` and `getOneSolution`.

Let us first describe the syntax of our core language. We will use variables taken from a finite set $Var = \{x_1, \ldots, x_m\}$, for simplicity all of type integer ($m \in \mathbb{N}$). Also let $Op = AOp \cup BOp \cup ROp = \{+, -, *, /\} \cup \{\&\&, ||\} \cup \{==, !=, <=, >=, <, >\}$ be a finite set of arithmetic, boolean, and relational operation symbols, respectively. We focus on binary operation symbols. Furthermore, $\mathcal{M}$ is a finite set of methods.[3]

The syntax of arithmetic expressions and boolean expressions as well as statements can be described by the following grammar. $AExpr$, $BExpr$, and $Stat$ denote the sets of all arithmetic expressions, boolean expressions, and statements, respectively, which can be constructed by the rules of this grammar.

---

[2] Nevertheless, Muli's symbolic JVM supports these features exactly according to the JVM specification [12] (but does not add interesting details w.r.t. non-determinism).

[3] In fact they are functions, since we ignore object-orientation in this presentation.

$$e ::= c \mid x \mid e_1 \oplus e_2 \mid m(e_1, \ldots, e_k)$$
$$\text{where } c \in \mathbb{Z}, \ x \in Var, \ e_1, \ldots, e_k \in AExpr, \ \oplus \in AOp, \ m \in \mathcal{M}, \ k \in \mathbb{N},$$

$$b ::= e_1 \odot e_2 \mid b_1 \otimes b_2 \mid \text{true} \mid \text{false}$$
$$\text{where } e_1, e_2 \in AExpr, \ b_1, b_2 \in BExpr, \ \odot \in ROp, \ \otimes \in BOp,$$

$$s ::= ; \mid \text{int } x; \mid \text{int } x \text{ free}; \mid x = e; \mid e; \mid \{s\} \mid s_1 \ s_2 \mid$$
$$\text{if } (b) \ s_1 \text{ else } s_2 \mid \text{while } (b) \ s \mid \text{return } e; \mid \text{fail};$$
$$\text{where } x \in Var, \ e \in AExpr, \ b \in BExpr, \ s, s_1, s_2 \in Stat.$$

Note, in particular, the possibility to create free logic variables by int x free;.

After describing the syntax of the core language, let us now define its semantics. In the sequel, let $\mathcal{A} = \{\alpha_0, \ldots, \alpha_n\}$ be a finite set of memory addresses ($n \in \mathbb{N}$). Moreover, let

$$Tree(\mathcal{A}, \mathbb{Z}) = \mathcal{A} \cup \mathbb{Z} \cup \{\oplus(t_1, t_2) \mid t_1, t_2 \in Tree(\mathcal{A}, \mathbb{Z}), \oplus \in Op\}$$

be the set of all symbolic expression trees with addresses and integer constants as leaves and operation symbols as internal nodes.

We provide a reduction semantics, where the computations depend on an environment, a state, and a constraint store. Let $Env = (Var \cup \mathcal{M}) \to (\mathcal{A} \cup (Var^* \times Stat))$ be the set of all environments, mapping each variable to an address and each function to a representation $((x_1, \ldots, x_k), s)$ that describes its parameters and code, with the additional restriction that elements of Env may neither map variables to parameters and code nor functions to addresses. We consider functions to be in global scope and define a special initial environment $\rho_0 \in Env$ that maps functions to their respective parameters and code. Moreover, let $\Sigma = \mathcal{A} \to (\{\bot\} \cup Tree(\mathcal{A}, \mathbb{Z}))$ be the set of all possible memory states. In $\sigma \in \Sigma$, a special address α_0 with $\sigma(\alpha_0) = \bot$ is reserved for holding return values of method invocations. Furthermore, $CS = \{\text{true}\} \cup Tree(\mathcal{A}, \mathbb{Z})$ is the set of all possible constraint store states. Since constraints are specific boolean expressions, only conjunctions and relational operation symbols such as $==$ and $>$ will appear at the root of such a tree.

In the sequel, $\rho \in Env, \sigma \in \Sigma, \gamma \in CS$; if needed, we will also add discriminating indices. We will use the notation $a[x/d]$ when modifying a state or environment a, meaning

$$a[x/d](b) = \begin{cases} d & \text{, if } b = x \\ a(b) & \text{, otherwise.} \end{cases}$$

A free variable is represented by a reference to its own location in memory. Consequently, $\sigma(\rho(x)) = \rho(x)$ if x is a free variable. Initially, a constraint store γ is empty, i.e. it is initialised with true. During execution of a program, constraints may incrementally be added to the store. This is done by imposing a conjunction of the existing constraints and a new constraint, thus replacing the constraint store by the new conjunction. As a result, the constraint store is typically described by a conjunction of atomic boolean expressions. We treat the constraint solver as a black box. In our implementation, we use the external constraint solver JaCoP [11] in its most recent version 4.4. In fact, the constraint solver is exchangeable and any solver implementation fulfilling our requirements (particularly incremental adding/removal of constraints) can be used.[4]

Note that our definition of functions does not fully cover the concept of methods in object-oriented languages, since we abstract from classes and, therefore, inheritance. However, a function in our semantics can be compared to a static method, since a function in this semantics can access and modify its own arguments and variables, but not instance variables of an object. Static fields could be modelled as global variables, i.e. further entries in ρ_0.

Since classes, inheritance, instance variables, and static variables have little influence on the interaction between imperative statements, free variables, and non-determinism, object orientation can be considered (almost) orthogonal to our work.

3.1   Semantics of Expressions

Let us start with the semantics of expressions. The semantics of expressions is described by a relation $\rightarrow \subset (Expr \times Env \times \Sigma \times CS) \times ((\mathbb{B} \cup Tree(\mathcal{A}, \mathbb{Z})) \times \Sigma \times CS)$, which we use in infix notation. Note that evaluating an expression can, in general, change state and constraint store as a side effect, although only the Invoke rule actively does so. We will point out expressions that make use of this, whereas the others merely propagate changes (if any) resulting from the evaluation of subexpressions.

The treatment of constants and variables is trivial.

$$\langle c, \rho, \sigma, \gamma \rangle \rightarrow (c, \sigma, \gamma), \text{ if } c \in \mathbb{Z} \cup \mathbb{B} \qquad \text{(Con)}$$

$$\langle x, \rho, \sigma, \gamma \rangle \rightarrow (\sigma(\rho(x)), \sigma, \gamma) \qquad \text{(Var)}$$

Nested arithmetic expressions without free variables are evaluated directly, whereas expressions comprising free variables result in a (deterministic) unevaluated (!) symbolic expression ($\in Tree(\mathcal{A}, \mathbb{Z})$).

$$\frac{\langle e_1, \rho, \sigma, \gamma \rangle \rightarrow (v_1, \sigma_1, \gamma_1), \quad \langle e_2, \rho, \sigma_1, \gamma_1 \rangle \rightarrow (v_2, \sigma_2, \gamma_2), \quad v_1, v_2, v = v_1 \oplus v_2 \in \mathbb{Z}}{\langle e_1 \oplus e_2, \rho, \sigma, \gamma \rangle \rightarrow (v, \sigma_2, \gamma_2)} \quad \text{(AOp1)}$$

$$\frac{\langle e_1, \rho, \sigma, \gamma \rangle \rightarrow (v_1, \sigma_1, \gamma_1), \quad \langle e_2, \rho, \sigma_1, \gamma_1 \rangle \rightarrow (v_2, \sigma_2, \gamma_2), \quad \{v_1, v_2\} \not\subseteq \mathbb{Z}}{\langle e_1 \oplus e_2, \rho, \sigma, \gamma \rangle \rightarrow (\oplus(v_1, v_2), \sigma_2, \gamma_2)} \quad \text{(AOp2)}$$

A boolean expression of the form $e_1 \odot e_2$ is evaluated analogously.

---

[4] A very simple constraint solver could just take equality constraints into account. In this case, $\gamma \models x == v$, if $\gamma = b_1 \wedge \ldots \wedge b_k$ and for some $j \in \{1, \ldots, k\}$ $b_k = (x == v)$.

Coherent with Java, conjunctions of boolean expressions are evaluated non-strictly. The rules for the non-strict boolean disjunction operator $||$ are defined analogously to the following rules for &&.

$$\frac{\langle b_1, \rho, \sigma, \gamma \rangle \rightarrow (v_1, \sigma_1, \gamma_1), \quad \gamma \models \neg v_1}{\langle b_1 \text{ \&\& } b_2, \rho, \sigma, \gamma \rangle \rightarrow (\texttt{false}, \sigma_1, \gamma_1)} \tag{And1}$$

$$\frac{\langle b_1, \rho, \sigma, \gamma \rangle \rightarrow (v_1, \sigma_1, \gamma_1), \quad \gamma \not\models \neg v_1, \quad (b_2, \sigma_1, \gamma_1) \rightarrow (v_2, \sigma_2, \gamma_2)}{\langle b_1 \text{ \&\& } b_2, \rho, \sigma, \gamma \rangle \rightarrow (\wedge(v_1, v_2), \sigma_2, \gamma_2)} \tag{And2}$$

We consider a function invocation to be an expression as well, as the caller can use its result in a surrounding expression. Evaluation of the function is likely to result in a state change as well as in additions to the constraint store. Invoking m implies that its description $\rho(m)$ is looked up and corresponding fresh addresses $\alpha_1, \ldots, \alpha_k$, one for each of its k parameters, are created. The corresponding memory locations are initialised by the caller. Note that the respective values can contain free variables. $\sigma_{k+1}(\alpha_0)$ will contain the return value from evaluating the $\texttt{return}$ statement in the body, whose semantics will be defined later (cf. rule Ret). As the compiler enforces the presence of a $\texttt{return}$ statement, we can safely assume that $\sigma_{k+1}(\alpha_0)$ holds a value after reducing s. Invoke resets that value to $\bot$ for further evaluations within the calling method. We use the shorthand notation $\bar{a}_k = (a_1, \ldots, a_k)$ for vectors of k elements.

$$\frac{\begin{array}{c}\langle e_1, \rho, \sigma, \gamma \rangle \rightarrow (v_1, \sigma_1, \gamma_1), \quad \langle e_2, \rho, \sigma_1, \gamma_1 \rangle \rightarrow (v_2, \sigma_2, \gamma_2), \ldots, \\ \langle e_k, \rho, \sigma_{k-1}, \gamma_{k-1} \rangle \rightarrow (v_k, \sigma_k, \gamma_k), \quad \rho(m) = (\bar{x}_k, \ s), \\ \langle s, \rho_0[\bar{x}_k/\bar{\alpha}_k], \sigma_k[\bar{\alpha}_k/\bar{v}_k], \gamma_k \rangle \rightsquigarrow (\rho_{k+1}, \sigma_{k+1}, \gamma_{k+1}), \quad \sigma_{k+1}(\alpha_0) = r \end{array}}{\langle m(e_1, \ldots, e_k), \rho, \sigma, \gamma \rangle \ \rightarrow (r, \sigma_{k+1}[\alpha_0/\bot], \gamma_{k+1})} \tag{Invoke}$$

3.2   Semantics of Statements

Next, we describe the semantics of statements by a relation $\rightsquigarrow \subset (Stat \times Env \times \Sigma \times CS) \times (Env \times \Sigma \times CS)$, which we also use in infix notation.

A variable declaration changes the environment by reserving a fresh memory location α for that variable. A free variable is represented by a reference to its own location. Enclosing declarations in a block ensures that changes of the environment stay local.

$$\langle \texttt{int } x;, \rho, \sigma, \gamma \rangle \rightsquigarrow (\rho[x/\alpha], \sigma, \gamma) \tag{Decl}$$

$$\langle \texttt{int } x \texttt{ free};, \rho, \sigma, \gamma \rangle \rightsquigarrow (\rho[x/\alpha], \sigma[\alpha/\alpha], \gamma) \tag{Free}$$

$$\frac{\langle s, \rho, \sigma, \gamma \rangle \rightsquigarrow (\rho_1, \sigma_1, \gamma_1)}{\langle \{ \ s \ \}, \rho, \sigma, \gamma \rangle \rightsquigarrow (\rho, \sigma_1, \gamma_1)} \tag{Block}$$

As a particularity of a constraint-logic OO language, an assignment $x = e$ cannot just overwrite a location in memory corresponding to x, since this might have an unwanted side effect on constraints that involve x and refer to its former value. This side effect might turn such constraints unsatisfiable after they have been imposed and checked, thus leaving a currently executed branch in an inconsistent state. We avoid this by assigning a new memory address α_1 to the variable on the left-hand side. At the new address, we store the result from evaluating the right-hand side. Consequently, old constraints or expressions that involve the former value of x are deliberately left untouched by the assignment. In contrast, later uses of the variable refer to its new value. The environment is updated to achieve this behaviour.

$$\frac{\langle e, \rho, \sigma, \gamma \rangle \rightarrow (v, \sigma_1, \gamma_1)}{\langle x = e, \rho, \sigma, \gamma \rangle \rightsquigarrow (\rho[x/\alpha_1], \sigma_1[\alpha_1/v], \gamma_1)} \quad \text{(Assign)}$$

Since the syntax does not enforce that no statements follow a `return` statement, we provide sequence rules that take into account that the state may hold a value in α_0 (indicating a preceding `return`) or not ($\perp$). Further statements are executed iff the latter is the case. Otherwise, further statements are discarded as a preceding `return` has already provided a result in α_0.

$$\frac{\langle s_1, \rho, \sigma, \gamma \rangle \rightsquigarrow (\rho_1, \sigma_1, \gamma_1), \quad \sigma_1(\alpha_0) == \perp,}{\langle s_1 \ s_2, \rho, \sigma, \gamma \rangle \rightsquigarrow (\rho_2, \sigma_2, \gamma_2)} \quad \text{(Seq)}$$

$$\frac{\langle s_1, \rho, \sigma, \gamma \rangle \rightsquigarrow (\rho_1, \sigma_1, \gamma_1), \quad \sigma_1(\alpha_0) \neq \perp}{\langle s_1 \ s_2, \rho, \sigma, \gamma \rangle \rightsquigarrow (\rho_1, \sigma_1, \gamma_1)} \quad \text{(SeqFin)}$$

The two following rules for if-statements introduce non-determinism in case that the constraints neither entail the branching condition nor its negation.[5]

$$\frac{\langle b, \rho, \sigma, \gamma \rangle \rightarrow (v, \sigma_1, \gamma_1), \quad \gamma_1 \not\models \neg v, \quad \langle s_1, \rho, \sigma_1, \gamma_1 \wedge v \rangle \rightsquigarrow (\rho_1, \sigma_2, \gamma_2)}{\langle \text{if } (b) \ s_1 \text{ else } s_2, \rho, \sigma, \gamma \rangle \rightsquigarrow (\rho_1, \sigma_2, \gamma_2)} \quad \text{(If}_t)$$

$$\frac{\langle b, \rho, \sigma, \gamma \rangle \rightarrow (v, \sigma_1, \gamma_1), \quad \gamma_1 \not\models v, \quad \langle s_2, \rho, \sigma_1, \gamma_1 \wedge \neg v \rangle \rightsquigarrow (\rho_1, \sigma_2, \gamma_2)}{\langle \text{if } (b) \ s_1 \text{ else } s_2, \rho, \sigma, \gamma \rangle \rightsquigarrow (\rho_1, \sigma_2, \gamma_2)} \quad \text{(If}_f)$$

As with `if`, `while` can also behave non-deterministically.

$$\frac{\langle b, \rho, \sigma, \gamma \rangle \rightarrow (v, \sigma_1, \gamma_1), \quad \gamma_1 \not\models \neg v, \quad \langle s, \rho, \sigma_1, \gamma_1 \wedge v \rangle \rightsquigarrow}{(\rho_1, \sigma_2, \gamma_2), \ \langle \text{while } (b) \ s, \rho_1, \sigma_2, \gamma_2 \rangle \rightsquigarrow (\rho_2, \sigma_3, \gamma_3)}{\langle \text{while } (b) \ s, \rho, \sigma, \gamma \rangle \rightsquigarrow (\rho_2, \sigma_3, \gamma_3)} \quad \text{(Wh}_t)$$

$$\frac{\langle b, \rho, \sigma, \gamma \rangle \rightarrow (v, \sigma_1, \gamma_1), \quad \gamma_1 \not\models v}{\langle \text{while } (b) \ s, \rho, \sigma, \gamma \rangle \rightsquigarrow (\rho, \sigma_1, \gamma_1 \wedge \neg v)} \quad \text{(Wh}_f)$$

---

[5] In the implementation, the applicability of these rules will depend on the constraint propagation abilities of the employed constraint solver. We discuss the implications in Sect. 5.

All branching rules If_f, If_t, Wh_f, and Wh_t could be accompanied by more efficient ones that deterministically choose a branch if its condition does not involve free variables, i.e. without having to consult the constraint store. We omit these rules in an effort to keep our definitions concise, as the provided ones can also handle these cases.

We assume that the code of a user-defined function is terminated by a `return` statement, i.e. its existence has to be ensured by the compiler. The corresponding return value is supplied to the caller by storing it in α_0, causing remaining statements of the function to be skipped (cf. rule SeqFin), and letting the caller extract the result from α_0 (cf. rule Invoke). The `return` statement is handled as follows:

$$\frac{\langle e, \rho, \sigma, \gamma \rangle \rightarrow (v, \sigma_1, \gamma_1)}{\langle \texttt{return } e, \rho, \sigma, \gamma \rangle \rightsquigarrow (\rho, \sigma_1[\alpha_0/v], \gamma_1)} \tag{Ret}$$

Furthermore, we do not define an evaluation rule involving a `fail` statement. This is intentional, as the evaluation of such a statement leads to a computation that fails immediately.

The following (optional) substitution rule allows to simplify expressions and results.

$$\frac{\gamma \models \gamma(\alpha) == v, \quad \langle s, \rho, \sigma[\alpha/v], \gamma \rangle \rightsquigarrow (\rho_1, \sigma_1, \gamma_1)}{\langle s, \rho, \sigma, \gamma \rangle \rightsquigarrow (\rho_1, \sigma_1, \gamma_1)} \tag{Subst}$$

When variables are not sufficiently constrained to concrete values, *labeling* can be used to substitute variables for values that satisfy the imposed constraints [5]. This non-deterministic rule is applied with the least priority, i.e. it should only be used if no other rule can be applied. Otherwise, it would result in a lot of non-deterministic branching, thus preventing the constraint solver from an efficient reduction of the search space by constraint propagation.

$$\frac{\gamma \not\models \sigma(\alpha) \neq v, \quad \langle s, \rho, \sigma[\alpha/v], \gamma \wedge (\sigma(\alpha) == v) \rangle \rightsquigarrow (\rho_1, \sigma_1, \gamma_1)}{\langle s, \rho, \sigma, \gamma \rangle \rightsquigarrow (\rho_1, \sigma_1, \gamma_1)} \tag{Label}$$

4   Example Evaluation

We demonstrate the use of the reduction rules defined in Sect. 3 by computing one possible result of the logarithm program from Listing 1 that will be invoked by an additional method **int** main() { **return** log(1); }. Other possible results can be computed analogously. We abbreviate the code of log and pow by s_1 and s_3, respectively, to improve readability. The substatement s_2 is included in s_1, while s_3 includes the substatements s_4, s_5, and s_6. Moreover, we use the infix notation for nested expressions, e.g. we write $n \geq 1$ instead of $\geq (n, 1)$.

Initially, let $\rho_0 = \{main \mapsto (\epsilon, \texttt{return log(1);}), \; log \mapsto ((x), s_1), \; pow \mapsto ((b, y), s_3)\}$. Furthermore, let $\gamma_1 = true$ and $\sigma_0 = \{\alpha_0 \mapsto \bot\}$. We begin in method $\texttt{main()}$, which evaluates to

$$\frac{\dfrac{\langle 1, \rho_0, \sigma_0, \gamma_1 \rangle \to (1, \sigma_0, \gamma_1) \; (\text{Con}), \; \rho_0(log) = ((x), s_1),}{\dfrac{(\text{Lemma}_1), \; \sigma_6(\alpha_0) = 0}{\langle \texttt{log(1)}, \rho_0, \sigma_0, \gamma_1 \rangle \to (0, \sigma_6[\alpha_0/\bot], \alpha_2 == 0)} \; (\text{Invoke})}}{\langle \texttt{return log(1)}, \rho_0, \sigma_0, \gamma_1 \rangle \rightsquigarrow (\rho_0, \sigma_6[\alpha_0/0], \alpha_2 == 0)} \; (\text{Ret})$$

Performing an entire evaluation with this example is interesting, but lengthy. We therefore moved the detailed evaluation into the appendix (cf. Lemma$_1$) and use the opportunity to highlight some interesting evaluation steps here. In the final state, $\sigma_6 = \sigma_0[\alpha_0/0, \alpha_1/1, \alpha_2/\alpha_2, \alpha_3/2, \alpha_4/\alpha_2, \alpha_7/0, \alpha_8/1]$.

The final result $\sigma_6(\alpha_0) = 0$ results from the constraint $\alpha_2 \leq 0$ obtained from evaluating Wh_f (Lemma$_9$ in the appendix provides context):

$$\frac{\dfrac{\langle \texttt{i}, \rho_4, \sigma_3, \gamma_1 \rangle \to (0, \sigma_3, \gamma_1) \; (\text{Var}),}{\dfrac{\langle \texttt{y}, \rho_4, \sigma_3, \gamma_1 \rangle \to (\alpha_2, \sigma_3, \gamma_1) \; (\text{Var})}{\langle \texttt{i < y}, \rho_4, \sigma_3, \gamma_1 \rangle \to (0 < \alpha_2, \sigma_3, \gamma_1)} \; (\text{AOp2}),}{\dfrac{\gamma \not\models (0 < \alpha_2)}{\langle \texttt{while (i < y) } s_6, \rho_4, \sigma_3, \gamma_1 \rangle \rightsquigarrow (\rho_4, \sigma_3, \gamma_1 \wedge \neg(0 < \alpha_2))} \; (\text{Wh}_f)}$$

where $\rho_4 = \rho_0[b/\alpha_3, y/\alpha_4, i/\alpha_7, r/\alpha_8]$ and $\sigma_0[\alpha_1/1, \alpha_2/\alpha_2, \alpha_3/2, \alpha_4/\alpha_2, \alpha_7/0, \alpha_8/1]$. $\alpha_2 \leq 0$ is further refined to $\alpha_2 == 0$ by the labeling rule in Lemma$_2$ in the appendix.

In Lemma$_2$, the constraint store is used to deduce that $\alpha_2 == 0$ is consistent with the current constraint, $\alpha_2 \leq 0$, as well as with the constraint store γ_2. Therefore, labeling non-deterministically imposes the more restrictive constraint $\alpha_2 == 0$. Other branches may impose further constraints consistent with $\alpha_2 < 0$.

If we had non-deterministically chosen rule Wh_t in Lemma$_9$, we would have performed an iteration of the $\texttt{while}$ loop, leading to more computations that would not result in solutions, as they would be discarded as incorrect by the $\texttt{fail}$ statement of the $\texttt{log}$ method.

The evaluation of rule Assign in Lemma$_7$ creates a new memory location α_7 in σ_2 for the new value of $\texttt{i}$ and updates the environment accordingly. At this point, no references to the old location α_5 exist, so an implementation could use garbage collection to free that location. Hypothetically, if rule Wh_t had been chosen in Lemma$_9$, an iteration of the loop would have resulted in additional evaluations of rule Assign, e.g. to increment $\texttt{i}$, thus reserving additional locations. In the case of $\texttt{i}$, the new value would depend on the value in α_7. However, as the old value and the increment are constant, the new value would be computed by evaluating rule AOp1, so that, again, no reference to α_7 is needed.

5   Discussion

The key aspect of the semantic rules for the presented core language is the interaction between constraint-logic programming and imperative programming. Some aspects of it offer themselves for thorough discussion.

The (potentially) non-deterministic evaluation of our rules If_f, If_t, Wh_f, and Wh_t highly depends on the included constraint solver. Our definition allows to follow a branch if the negation of its condition is not entailed by the current constraint store γ. When implementing this, a constraint solver will be used to check whether $\gamma \not\models \neg v$ (analogously for $\gamma \not\models v$). If the constraint solver is not able to show that the constraints entail $\neg v$, this may have three reasons: (1) $\gamma \models v$, or (2) the current constraints neither entail v nor $\neg v$, or (3) the constraint propagation abilities of the employed constraint solver are insufficient to show that $\gamma \models \neg v$, but in fact $\gamma \models \neg v$. In case (1), the system behaves deterministically and only one rule for if (or while) will be applied. In case (2), one of the two rules for if (or while) can be chosen non-deterministically. Only case (3) is problematic. In this case, a branch can be chosen that corresponds to inconsistent constraints. In practice, solvers do not achieve perfect constraint propagation and also no global consistency of the constraints. Consequently, results corresponding to inconsistent constraints may only be discovered later, e.g. during labeling. In the meantime, non-backtrackable statements (e.g. ones that result in input/output) of search regions may have been executed in branches that prove infeasible later. Thus, we suggest to avoid input/output in search regions.

We would like to point out that the aforementioned problem is not specific to Muli, as this can occur in Prolog (using CLP(FD) [20]) as well. Consider the Prolog program provided in Listing 2. When you execute the first goal, the output will (among the unreduced constraint system) contain a line that says successful, even though it is apparent to the human reader that there is no solution, so that the write statement should not have been reached. In contrast, if label is invoked before write (second goal), Prolog realises that there is no solution and therefore gives the correct result false.

```
use_module(library(clpfd)).
?- [X,Y,Z] ins 0..1, all_different([X,Y,Z]),
     write('successful').
?- [X,Y,Z] ins 0..1, all_different([X,Y,Z]),
     label([X,Y,Z]), write('successful').
```

Listing 2. Demonstration of the limits of constraint propagation using an example in Prolog+CLP(FD).

We see two options to handle this situation in Muli programs. The first option is to explicitly label variables sufficiently at every branch such that the constraint solver is able to either infer $\gamma \models v$ or $\gamma \models \neg v$. However, as explained in context of the Label rule, this also introduces a lot of non-deterministic branching by creating one branch per label. Therefore, the effectivity of constraint propagation is reduced and the overall effort for search is increased. For the same reason we decided that Muli should not implicitly perform labeling at every branch either, as performance would deteriorate.

The second option is to perform labeling only after a solution has been found during encapsulated search. In fact, such a solution is merely a potential solution, under the condition that the corresponding constraints are also satisfiable. As a result, encapsulated search produces a stream of pairs, each of which comprises one potential solution and its corresponding set of constraints. Thus, at this point the enclosing application can iterate over this stream and perform (sufficient) labeling, until it is clear whether the constraints are actually satisfiable. This rules out infeasible solutions afterwards. The implementation of Muli provides an explicit label operation, which the application developer can use for this purpose. We decided not to do this implicitly in order to give the developer more flexibility. It is easy to wrap this functionality into a search operation which labels every found solution implicitly.

Both mentioned options are available to the developer. We recommend the second one, possibly in the wrapped version with implicit labeling. For search regions that involve only backtrackable statements, the result does not depend on the chosen option, but the second option is presumably more efficient as fewer branches have to be evaluated. For other search regions, only the first option can avoid unwanted side effects of illegally accessed branches. However, search then becomes less efficient. Therefore, in case that non-backtrackable side effects have to be avoided, we recommend that the developer removes input/output operations from search regions and moves them behind encapsulation instead.

Formalising the operational semantics of Muli has also helped uncover some operations whose semantics are sufficiently clear in deterministic Java, but become ambiguous when non-determinism and symbolic execution are added. Consequently, some alternatives could be discussed on a conceptual level using this semantics, before deriving a corresponding implementation. This particularly involves the interpretation of symbolic variables (rules Invoke and Var) and assignments, as outlined subsequently.

By rule Assign, an assignment x = e creates a new memory address for the variable x and changes the environment accordingly. As a result, memory usage of a Muli program is increased with every assignment, instead of with every declaration of a variable as in imperative OO languages. Nevertheless, this behaviour is required in order to avoid unwanted side effects on previous constraints involving x. The alternative, mutating $\sigma(\rho(x))$ directly, would result in assignments to x that could render constraints involving x unsatisfiable ex post, i.e. after branching has occurred that depended on such a constraint.

As another consequence, rule Assign ensures that the interpretation of symbolic variables is equivalent to that of regular values. Consider the simple excerpt from a Java program given in Listing 3 as an example: After evaluating the last line, y is still expected to be 5, even though x now holds a different value. After all, although primitive variables can be directly mutated in Java, their previous interpretations cannot. Similarly, for symbolic values, rule Assign ensures that references before and after an assignment are treated distinctly, even though memory efficiency is adversely affected. Nevertheless, unreferenced former meanings of a variable may be destroyed by the garbage collector, thus reclaiming (some) memory.

```
int x = 5; int y = x;
x = 3;
```

Listing 3. Minimal example demonstrating that variables may be mutated directly, in contrast to results of their uses: After evaluation, y is 5.

Implicitly, our rules Assign (or Invoke) and Var enable sharing of symbolic values. Assigning a free variable x to another free variable y means that the address $\rho(x)$ of x is stored in the memory location corresponding to y by modifying state as $\sigma[\rho(y)/\rho(x)]$. Consequently, subsequent constraints and expressions that involve either variable will actually reference the same variable. The sharing behaviour is exhibited in the example in Lemma$_5$ in the appendix, where a free variable is passed to the pow method as its second parameter. pow adds constraints to that variable that only come into effect when labeling is performed in its invoking context in log (Lemma$_2$ in the appendix).

Regarding backtracking, the implementation is only implicitly affected by the presented operational semantics. Here, the semantics defines the desired state of the overall VM that must be achieved before evaluation in terms of $\rho \in Env$, $\sigma \in \Sigma$, $\gamma \in CS$. Considering the multitude of options for achieving the desired VM state that lend themselves for the implementation, we briefly outline the options without prescribing either. Firstly, "don't care" non-determinism considers only one evaluation alternative and therefore does not require backtracking at all. Secondly, it would be possible to fork at statements that introduce non-determinism, thus evaluating all alternatives in parallel. This does not require backtracking either, however, consider that this generates a lot of overhead in terms of memory and computation, as the VM must be forked in its entirety to accommodate for any side effects, and as all forks must be joined in order to return to deterministic computation after a search region is fully processed. Thirdly, the alternatives can be evaluated sequentially. To achieve this, the VM must record changes to the data structures on a trail equivalent to that of Prolog in order to reconstruct a previous state during backtracking. Our implementation resorts to the latter option using a trail adapted from the Warren Abstract Machine. Nevertheless, the remaining options would also be interesting to pursue.

6 Related Work

To the best of our knowledge, this paper is the first to present a formal semantics of an imperative language enhanced by features of constraint-logic programming. For sake of clarity we focused on a core language. A full formal semantics of Java alone may require an entire book as in the work by Stärk et al. [19]. K-Java [2] is another approach to define a formal semantics of Java. However, in the cited paper the authors focus on selected aspects of the language. The official semantics of Java is extensively described in natural language (cf. [6,12]).

Some existing core languages of Java such as Featherweight Java [9] are tailored to the investigation of the typing system and not meant to be executable. Hainry [7] investigates an object-oriented core language focussing on computational complexity. As a result of their respective foci they were not suitable to be extended for Muli.

The encapsulated search of Muli has been inspired and adapted from the corresponding feature of the functional-logic language Curry. An operational semantics of Curry can be found in [1]. It is simpler than our semantics, since Curry is purely declarative and does not have to bother with side effects.

Approaches for integrating object-oriented features into a (constrained) logic language are e.g. Oz [21], Visual Prolog [17], Prolog++ [15], and Concurrent Prolog [18]. However, these approaches maintain a declarative flavour and mainly provide syntactic sugar for object-orientation. They are unfamiliar for mainstream object-oriented programmers.

There are also approaches which add constrained-logic features to an imperative/object-oriented language. Typically, the integration is less seamless than in Muli and the language parts stemming from different paradigms can clearly be distinguished [3,4]. CAPJa combines Java and Prolog and provides a simplified interface mapping Java objects to Prolog terms, but requires distinct code in each language nevertheless [16]. LogicJava [14] is more restrictive than Muli and only allows class fields to be logic variables. Moreover, entire methods have to be declared as searching or non-searching.

7   Conclusions and Future Work

Our work formalises an operational reduction semantics for an imperative core of the novel integrated constraint-logic object-oriented language Muli. Muli extends Java by logic variables, non-determinism, encapsulated search, and constraint solving. Muli is particularly suited for enterprise applications that involve both searching and non-searching business logic. Encapsulated search ensures that non-determinism is only introduced deliberately where needed, instead of spreading out over the whole program. Thus, the code outside of encapsulated search regions behaves just as ordinary Java code.

The presented operational semantics provides a basis for implementations of compiler, symbolic JVM, and tools for processing Muli programs. In particular, the formalisation has helped clarify possible ambiguities w.r.t. the semantics of certain statements under non-determinism, such as that of assignments to variables and uses of them. Furthermore, the semantics will facilitate reasoning about programs developed in Muli as demonstrated in the example evaluation. We made the symbolic JVM that executes Muli programs available as free software on GitHub.[6]

---

[6] https://github.com/wwu-pi/muli-env.

As future work, we would like to extend our core language and its semantics by more features of Java, such as classes and inheritance. We expect these additions to be quite orthogonal to the presently supported concepts. However, when (non-deterministically) instantiating a free variable with an object type, we have to take the whole corresponding inheritance hierarchy into account.

Appendix: Full Example Evaluation

In addition to ρ_0, σ_0, and γ_1 defined in section Sect. 4, the following auxiliary definitions will be needed as intermediate results: $\rho_1 = \rho_0[x/\alpha_1, y/\alpha_2]$, $\rho_2 = \rho_0[b/\alpha_3, y/\alpha_4]$, $\rho_3 = \rho_2[i/\alpha_5, r/\alpha_6]$, $\rho_4 = \rho_3[i/\alpha_7, r/\alpha_8]$, $\sigma_1 = \sigma_0[\alpha_1/1, \alpha_2/\alpha_2]$, $\sigma_2 = \sigma_1[\alpha_3/2, \alpha_4/\alpha_2]$, $\sigma_3 = \sigma_2[\alpha_7/0, \alpha_8/1]$, $\sigma_4 = \sigma_3[\alpha_0/1]$, $\sigma_5 = \sigma_4[\alpha_0/\bot]$, $\sigma_6 = \sigma_5[\alpha_0/0]$, $\gamma_2 = \gamma_1 \wedge \alpha_2 \leq 0$, and $\gamma_3 = \gamma_2 \wedge \alpha_2 == 0$. To simplify the understanding of the full computation provided in Sect. 4, we have decomposed it into a couple of lemmas. We present the computation in a top-down fashion. If you prefer a bottom-up fashion, just read the lemmas in reverse order. The names of the applied rules are specified in each step.

$$\frac{\langle \text{int y free;} \rangle \rightsquigarrow (\rho_0[x/\alpha_1, y/\alpha_2], \sigma_0[\alpha_1/1, \alpha_2/\alpha_2], \gamma_1) \; (\text{Free}),}{\langle \text{int y free; } s_2, \rho_0[x/\alpha_1], \sigma_0[\alpha_1/1], \gamma_1 \rangle \rightsquigarrow (\rho_1, \sigma_6, \gamma_3)} \; (\text{Seq})$$

(Lemma$_2$)

(Lemma$_1$)

$$\frac{(\text{Lemma}_3), \; \gamma_2 \nvDash \neg\text{true}, \; \dfrac{\gamma_2 \nvDash \sigma_5(\alpha_2) \neq 0, \; (\text{Lemma}_4)}{\langle \text{return y;}, \rho_1, \sigma_5, \gamma_2 \rangle \rightsquigarrow (\rho_1, \sigma_6, \gamma_3)} \; (\text{Label})}{\langle s_2, \rho_1, \sigma_1, \gamma_1 \rangle \rightsquigarrow (\rho_1, \sigma_6, \gamma_3)} \; (\text{If}_t)$$

(Lemma$_2$)

$$\frac{(\text{Lemma}_5), \; \langle x, \rho_1, \sigma_5, \gamma_2 \rangle \rightarrow (1, \sigma_5, \gamma_2) \; (\text{Var}), \; 1 == 1 = \text{true}}{\langle \text{pow(2, y)} == x, \rho_1, \sigma_1, \gamma_1 \rangle \rightarrow (\text{true}, \sigma_5, \gamma_2)} \; (\text{AOp1})$$

(Lemma$_3$)

$$\frac{\langle y, \rho_1, \sigma_5[\alpha_2/0], \gamma_2 \wedge \alpha_2 == 0 \rangle \rightarrow (0, \sigma_5, \gamma_3) \; (\text{Var})}{\langle \text{return y;}, \rho_1, \sigma_5, \gamma_2 \rangle \rightsquigarrow (\rho_1, \sigma_5[\alpha_0/0], \gamma_3)} \; (\text{Ret}) \qquad (\text{Lemma}_4)$$

$$\frac{\langle 2, \rho_1, \sigma_1, \gamma_1 \rangle \rightarrow (2, \sigma_1, \gamma_1) \; (\text{Con}),}{\langle y, \rho_1, \sigma_1, \gamma_1 \rangle \rightarrow (\alpha_2, \sigma_1, \gamma_1) \; (\text{Var}),}$$
$$\frac{\rho_1(pow) = ((b, y), s_3), \; (\text{Lemma}_6), \; \sigma_4(\alpha_0) = 1}{\langle \text{pow(2, y)}, \rho_1, \sigma_1, \gamma_1 \rangle \rightarrow (1, \sigma_4[\alpha_0/\bot], \gamma_2)} \; (\text{Invoke}) \qquad (\text{Lemma}_5)$$

$$\langle \texttt{int i;}, \rho_2, \sigma_2, \gamma_1 \rangle \rightsquigarrow (\rho_2[i/\alpha_5], \sigma_2, \gamma_1) \ (\text{Decl}),$$

$$\langle \texttt{int r;}, \rho_2[i/\alpha_5], \sigma_2, \gamma_1 \rangle \rightsquigarrow (\rho_2[i/\alpha_5, r/\alpha_6], \sigma_2, \gamma_1) \ (\text{Decl}),$$

$$\cfrac{\cfrac{(\text{Lemma}_7)}{\langle \texttt{int r; i = 0; r = 1; } s_4, \rho_2[i/\alpha_5], \sigma_2, \gamma_1 \rangle \rightsquigarrow (\rho_4, \sigma_4, \gamma_2)} \ (\text{Seq})}{\langle \texttt{int i; int r; i = 0; r = 1; } s_4, \rho_2, \sigma_2, \gamma_1 \rangle \rightsquigarrow (\rho_4, \sigma_4, \gamma_2)} \ (\text{Seq})$$

$$(\text{Lemma}_6)$$

$$\cfrac{\langle 0, \rho_3, \sigma_2, \gamma_1 \rangle \rightarrow (0, \sigma_2, \gamma_1) \ (\text{Con})}{\langle i = 0;, \rho_3, \sigma_2, \gamma_1 \rangle \rightsquigarrow (\rho_3[i/\alpha_7], \sigma_2[\alpha_7/0], \gamma_1)} \ (\text{Assign}),$$

$$\cfrac{\langle 1, \rho_3[i/\alpha_7], \sigma_2[\alpha_7/0], \gamma_1 \rangle \rightarrow (0, \sigma_2[\alpha_7/0], \gamma_1) \ (\text{Con})}{\langle r = 1;, \rho_3[i/\alpha_7], \sigma_2[\alpha_7/0], \gamma_1 \rangle \rightsquigarrow (\rho_4, \sigma_3, \gamma_1)} \ (\text{Assign}),$$

$$\cfrac{\cfrac{(\text{Lemma}_8)}{\langle r = 1; s_4, \rho_3[i/\alpha_7], \sigma_2[\alpha_7/0], \gamma_1 \rangle \rightsquigarrow (\rho_4, \sigma_4, \gamma_2)} \ (\text{Seq})}{\langle i = 0; r = 1; s_4, \rho_3, \sigma_2, \gamma_1 \rangle \rightsquigarrow (\rho_4, \sigma_4, \gamma_2)} \ (\text{Seq})$$

$$(\text{Lemma}_7)$$

$$\cfrac{(\text{Lemma}_9),}{\cfrac{\langle r, \rho_4, \sigma_3, \gamma_2 \rangle \rightarrow (1, \sigma_3, \gamma_2) \ (\text{Var})}{\langle \texttt{return r;}, \rho_4, \sigma_3, \gamma_2 \rangle \rightsquigarrow (\rho_4, \sigma_3[\alpha_0/1], \gamma_2)} \ (\text{Ret})}{\langle s_5; \texttt{ return r;}, \rho_4, \sigma_3, \gamma_1 \rangle \rightsquigarrow (\rho_4, \sigma_3[\alpha_0/1], \gamma_2)} \ (\text{Seq}) \quad (\text{Lemma}_8)$$

$$\langle i, \rho_4, \sigma_3, \gamma_1 \rangle \rightarrow (0, \sigma_3, \gamma_1) \ (\text{Var}),$$

$$\cfrac{\langle y, \rho_4, \sigma_3, \gamma_1 \rangle \rightarrow (\alpha_2, \sigma_3, \gamma_1) \ (\text{Var})}{\langle i < y, \rho_4, \sigma_3, \gamma_1 \rangle \rightarrow (0 < \alpha_2, \sigma_3, \gamma_1)} \ (\text{AOp2}),$$

$$\cfrac{\gamma \not\models (0 < \alpha_2)}{\langle \texttt{while (i < y) } s_6, \rho_4, \sigma_3, \gamma_1 \rangle \rightsquigarrow (\rho_4, \sigma_3, \gamma_1 \wedge \neg(0 < \alpha_2))} \ (\text{Wh}_f) \quad (\text{Lemma}_9)$$

References

1. Albert, E., Hanus, M., Huch, F., Oliver, J., Vidal, G.: An operational semantics for declarative multi-paradigm languages. Electron. Notes Theor. Comput. Sci. **70**(6), 65–86 (2002)
2. Bogdanas, D., Rosu, G.: K-Java: a complete semantics of Java. In: POPL 2015, pp. 1–12 (2015)
3. Cimadamore, M., Viroli, M.: A Prolog-oriented extension of Java programming based on generics and annotations. In: Amaral, V., et al. (eds.) Proceedings PPPJ, ACM ICPS, vol. 272, pp. 197–202. ACM (2007)
4. Cimadamore, M., Viroli, M.: Integrating Java and Prolog through generic methods and type inference. In: Wainwright, R.L., Haddad, H. (eds.) Proceedings of the 2008 SAC, pp. 198–205. ACM (2008). https://doi.org/10.1145/1363686
5. Frühwirth, T., Abdennadher, S.: Essentials of Constraint Programming. Springer, Heidelberg (2003). https://doi.org/10.1007/978-3-662-05138-2

6. Gosling, J., Joy, B., Steele, G., Bracha, G., Buckley, A.: The Java® Language Specification - Java SE 8 Edition (2015). https://docs.oracle.com/javase/specs/jls/se8/jls8.pdf

7. Hainry, E., Péchoux, R.: Objects in polynomial time. In: Feng, X., Park, S. (eds.) APLAS 2015. LNCS, vol. 9458, pp. 387–404. Springer, Cham (2015). https://doi.org/10.1007/978-3-319-26529-2_21

8. Hooker, J.N.: Operations research methods in constraint programming (Chap. 15). In: Rossi, F., van Beek, P., Walsh, T. (eds.) Handbook of CP. Elsevier, Amsterdam (2006)

9. Igarashi, A., Pierce, B.C., Wadler, P.: Featherweight Java: a minimal core calculus for Java and GJ. ACM Trans. Program. Lang. Syst. **23**(3), 396–450 (2001)

10. Kondoh, G., Onodera, T.: Finding bugs in Java native interface programs. In: ISSTA 2008, p. 109 (2008)

11. Kuchcinski, K.: Constraints-driven scheduling and resource assignment. ACM Trans. Des. Autom. Electron. Syst. **8**(3), 355–383 (2003)

12. Lindholm, T., Yellin, F., Bracha, G., Buckley, A.: The Java® Virtual Machine Specification - Java SE 8 Edition (2015). https://docs.oracle.com/javase/specs/jvms/se8/jvms8.pdf

13. Louden, K.C.: Programming Languages: Principles and Practice. Wadsworth Publ. Co., Belmont (1993)

14. Majchrzak, T.A., Kuchen, H.: Logic Java: combining object-oriented and logic programming. In: Kuchen, H. (ed.) WFLP 2011. LNCS, vol. 6816, pp. 122–137. Springer, Heidelberg (2011). https://doi.org/10.1007/978-3-642-22531-4_8

15. Moss, C.: Prolog++ - The Power of Object-Oriented and Logic Programming. International Series in Logic Programming. Addison-Wesley, Boston (1994)

16. Ostermayer, L.: Seamless cooperation of Java and Prolog for rule-based software development. In: Proceedings of the RuleML 2015, Berlin (2015)

17. Scott, R.: A Guide to Artificial Intelligence with Visual Prolog. Outskirts Press (2010)

18. Shapiro, E., Takeuchi, A.: Object oriented programming in concurrent Prolog. New Gener. Comput. **1**(1), 25–48 (1983)

19. Stärk, R., Schmid, J., Börger, E.: Java and the Java Virtual Machine - Definition, Verification, Validation. Springer, Heidelberg (2001). https://doi.org/10.1007/978-3-642-59495-3

20. Triska, M.: The finite domain constraint solver of SWI-Prolog. In: Schrijvers, T., Thiemann, P. (eds.) FLOPS 2012. LNCS, vol. 7294, pp. 307–316. Springer, Heidelberg (2012). https://doi.org/10.1007/978-3-642-29822-6_24

21. Van Roy, P., Brand, P., Duchier, D., Haridi, S., Schulte, C., Henz, M.: Logic programming in the context of multiparadigm programming: the Oz experience. Theory Pract. Log. Program. **3**(6), 717–763 (2003)

22. Warren, D.H.D.: An abstract Prolog instruction set. Technical report, SRI International, Menlo Park (1983)

Hypertree Decomposition: The First Step Towards Parallel Constraint Solving

Ke Liu[✉], Sven Löffler, and Petra Hofstedt

Brandenburg University of Technology Cottbus-Senftenberg, Cottbus, Germany
`liuke@b-tu.de`

Abstract. Parallel constraint solving is a promising way to enhance the performance of constraint programming. Yet, current solutions for parallel constraint solving ignore the importance of hypergraph decomposition when mapping constraints onto cores. This paper explains why and how the hypergraph decomposition can be employed to relatively evenly distribute workload in parallel constraint solving. We present our dedicated hypergraph decomposition method *det-k-CP* for parallel constraint solving. The result of *det-k-CP*, which conforms with four conditions of hypertree decomposition, can be used to allocate constraints of a given constraint network to cores for parallel constraint solving. Our benchmark evaluations have shown that *det-k-CP* can relatively evenly decompose a hypergraph for specific scale of constraint networks. Besides, we obtained competitive execution time as long as the hypergraphs are sufficiently simple.

Keywords: Parallel constraint solving · Hypertree decomposition

1 Introduction

Structural decomposition methods are one of research hot spots both in the area of relational databases and constraint programming. Many *NP-complete* and *NP-hard* problems can be solved in polynomial time if the corresponding hypergraph has the bounded hypertree-width, which indicates that the original intractable problem can be divided into a number of tractable subproblems [1]. In addition, the tree structure for a constraint network implies that each node of the tree decomposition can be solved simultaneously, which makes us naturally think of utilizing parallel computing to solve constraint satisfaction problem (CSP). In other words, the acyclic structure of constraint networks means that the given CSP problem is tractable and parallelizable [2–4]. Several decomposition methods have been developed to convert the cyclic constraint networks to acyclic ones although these methods apply to different types of graph for the given constraint network. For example, *join-tree-clustering* transforms the primal graph of the given constraint network into the equivalent acyclic network [5]. *Cycle-cutset decomposition* [6] also works on the primal graph by removing the vertexes that prevent the hypergraph to be acyclic. Some decomposition methods (e.g., *hinge*

D. Seipel et al. (Eds.): DECLARE 2017, LNAI 10997, pp. 81–94, 2018.
https://doi.org/10.1007/978-3-030-00801-7_6

decomposition [7], *hypertree decomposition* [1,8]), on the other hand, use the hypergraph as its input, and the output of these methods is at least in accord with the conditions for hypertree decomposition defined in [8].

Nevertheless, the decomposition methods, which have been proposed in the literature during the last decades, aim at obtaining as small hypertree width as possible for the hypergraph, because the smaller the width of a hypertree decomposition we obtained, the faster the original CSP problem can be solved [1]. Moreover, previous structural decomposition methods, such as *det-k-decomp* which is the most general decomposition method so far [9], cannot ensure a relatively even distribution of constraints based on our observation of results after running *det-k-decomp*. The algorithm *det-k-decomp* only guarantees the greatest node width of the decomposition tree is k, and fairly often, the width of most nodes is far less than k. This characteristic of *det-k-decomp* impedes its application in parallel constraint solving.

This paper intends to present a new decomposition *det-k-CP* method with stochastic search procedure for parallel constraint solving. The goal is to provide a mapping algorithm for parallel constraint solving. The idea behind *det-k-CP* is to utilize the property of dual graph that a redundant arc can be removed between two nodes of the graph if there is an alternative path that ensures two nodes still connected. We regard it as the first step towards parallel constraint solving due to the benefits that come from hypertree structure.

The rest of this paper is organized as follows. The basic definitions used in this study are presented in Sect. 2. Section 3 describes the new method in detail, and then analyses the time complexity of *det-k-CP*. In Sect. 4, we present our experimental results. Finally, we conclude in Sect. 5.

2   Preliminaries

A *constraint network* $\mathcal{R}$ is a triple (X, D, C), which consists of:

- a finite set of variables $X = \{x_1, \ldots, x_n\}$,
- a set of respective finite domains $D = \{D_1, \ldots, D_n\}$, where D_i is the domain of the variable x_i, and
- a set of constraints $C = \{c_1, \ldots, c_t\}$, where a constraint c_j is a relation R_j defined on a subset of variables S_j, $S_j \subseteq X$.

Any constraint network can be graphically represented by a *hypergraph*. A hypergraph $\mathcal{H}$ is a tuple (V, E), where V is a set of vertexes and E is a set of *hyperedges*. A hyperedge of a hypergraph is composed of an arbitrary number of vertexes, which makes hyperedges fundamentally different from normal edges in a graph. Any constraint in a given constraint network corresponds to a hyperedge in a hypergraph, and the variables of a constraint can be seen as vertexes of a hyperedge.

A *hypertree* of a hypergraph $\mathcal{H}$ is a triple (T, χ, λ), where $T = (V_T, E_T)$ is a tree, χ and λ are labeling functions. We denote a set of variables for a given node ($node_i$) in a hypertree by v_i. Therefore, $v_i = \chi(node_i)$ and

$v_i \subseteq 2^{vertexes(\mathcal{H})}$, where $vertexes(\mathcal{H})$ are vertexes of hypergraph $\mathcal{H}$. Similarly, we denote a set of edges of $node_i$ by e_i. Therefore, $e_i = \lambda(node_i)$ and $e_i \subseteq 2^{edges(\mathcal{H})}$, where $edges(\mathcal{H})$ are the hyperedges of hypergraph $\mathcal{H}$. By $root(T)$ we denote the root of a tree T, for every $p \in V_T$, let T_p denote the subtree of T with root p.

The *width of a hypertree* is the maximum number of hyperedges among the nodes of it, which is given by $hw(T) = max \mid \lambda(node_i) \mid$. *Hypertree decomposition* is a procedure that converts a hypergraph into a hypertree. In order to demonstrate hypertree decomposition on a given constraint network, assume we have a simple problem over a set of variables $\{x_1, \ldots, x_{10}\} \subseteq X$ modeled by the following constraints[1]

- *allDifferent1*(x_3, x_4, x_5, x_7)
- *allDifferent2*(x_1, x_4, x_6, x_9)
- *atLeastNvalues*(x_1, x_2, x_3)
- *arithm2*(x_6, x_8)

- *table1*(x_5, x_8, x_{10})
- *table2*(x_7, x_8, x_9)
- *arithm1*(x_5, x_6)

The hypergraph for this constraint network is depicted in Fig. 1, where the variables x_i, $i \in \{1, \ldots, 10\}$ are the vertexes, while the edges are represented by the enclosing ellipses. Figure 2 shows a possible hypertree decomposition of this hypergraph. Note that the hypertree decomposition (Fig. 2) can also be viewed as a *dual graph* for the hypergraph (Fig. 1). The nodes of a dual graph consist of a set of hyperedges of the corresponding hypergraph, and an edge of the dual graph is due to existing shared variables between two nodes of the dual graph. Formally, $\mathcal{H}^{dual}$ for a given $\mathcal{H}$ can be represented as a tuple (S, E) in which $S = \{s_1, ..., s_i, ..., s_j\} \subseteq edges(\mathcal{H})$ and $\forall e \in E = edges(\mathcal{H}^{dual}) : s_i \cap s_j = e \Leftrightarrow var(s_i) \cap var(s_j) \neq \emptyset$.

Gottlob *et al.* [1,8] defined four conditions, which must be satisfied by a hypertree after hypertree decomposition:

1. Every hyperedge of the hypergraph is contained in at least one node of the hypertree. This can be mathematically expressed as: $\forall e \in edges(H), \exists p \in vertexes(T) : e \subseteq \chi(p)$.
2. Nodes of a hypertree that contain the same vertex in the hypergraph form a subtree of the hypertree. For all $v \in vertices(H)$, the set

$$\{p \in vertexes(T) \mid v \in \chi(p)\}$$

   induces a connected subtree of T (This is also called connectedness property [5]).
3. For any node of the hypertree, the vertexes of χ are included in the vertexes of λ. $\forall p \in vertexes(T) : \chi(p) \subseteq vertexes(\cup\lambda(p))$.
4. For a vertex of a hypergraph in node p of the hypertree, if the vertex is included in both $\lambda(p)$ and $\chi(T_p)$, where $\chi(T_p)$ stands for the subtree of T rooted at p, this vertex must also be included in the $\chi(p)$. $\forall p \in vertexes(T) : vertexes(\cup\lambda(p)) \cap \chi(T_p) \subseteq \chi(p)$.

---

[1] The names of the constraints are consistent with the names of constraints used in the Choco Solver [10].

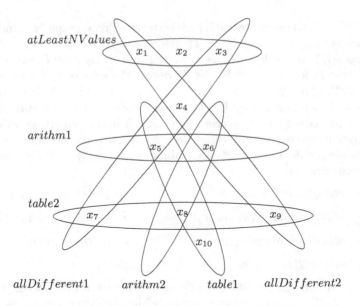

Fig. 1. The hypergraph for the constraint network. The example is based on [11]

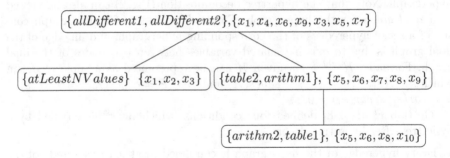

Fig. 2. Hypertree decomposition for the hypergraph of Fig. 1. The example is based on [11]

3   The Algorithm *det-k-CP*

In this section, we present our new algorithm *det-k-CP* which is designed to decompose a hypergraph for parallel constraint solving. The k of *det-k-CP* means the number of nodes in the decomposition tree. Roughly speaking, the target of *det-k-CP* is to decompose a given constraint network N to a degenerate tree in which each internal node has exactly one child so that the solutions of N can be found in time linear after each node is solved independently. Because *an acyclic constraint network can be solved efficiently* [5]. For example, Fig. 3 depicts a target degenerate decomposition tree with eight nodes decomposed by *det-k-CP* for a multi-core processor with eight cores.

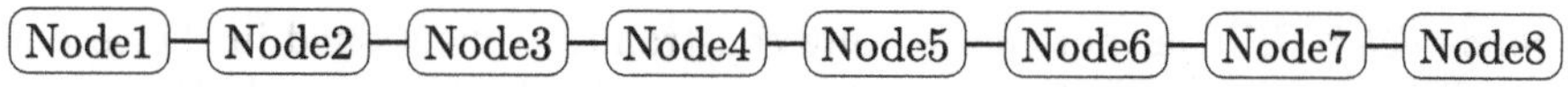

Node1 — Node2 — Node3 — Node4 — Node5 — Node6 — Node7 — Node8

Fig. 3. A constraint network is partitioned into eight parts. An edge between two nodes is due to the shared variables.

For a given ordering of nodes of a degenerate decomposition tree T generated by *det-k-CP*, there is an edge between two nodes because there exists shared variables between two nodes. Additionally, we only keep the edges between two adjacent nodes and eliminate the edges between any pair of non-adjacent nodes.[2] The mechanism of elimination of *det-k-CP*, which guarantees the decomposition tree is equivalent to the original one, is based on the property of which any edge on a circuit formed by common shared variables of a dual graph can be removed without changing the set of all solutions for the constraint network [5]. For instance, in Fig. 2, the edge between the root node and the right leaf node caused by shared variables (x_5, x_6) is inexistent because there are shared variables (x_5, x_6) between the root node and its child node, as well as the right leaf node and its parent node respectively. The positional relation between two nodes in T is either adjacent or non-adjacent; thus a pair of nodes in T can be denoted as (N_i, N_j) for non-adjacent or (N_p, N_{p+1}) for adjacent respectively, where $\mid i - j \mid \geq 2$, $0 \leq i < j \leq k$ and $p \in \{i, ..., j-1\}$. Mathematically, a decomposed graph T after decomposition by *det-k-CP* must meet the following two conditions:

$$\forall p \in \{i, ..., j-1\} : \chi(N_i) \cap \chi(N_j) \subseteq \chi(N_p) \cap \chi(N_{p+1}) \tag{1}$$

$$\forall N_i \in T : \cup \chi(N_i) = \chi(\mathcal{H}), and \cup \lambda(N_i) = \lambda(\mathcal{H}) \tag{2}$$

where the first condition means that the shared variables between any pair of nodes (N_p, N_{p+1}) in the interval $[i..j]$ must contain the shared variables between (N_i, N_j), whereas the second condition guarantees that the union of vertexes on each node of T is equal to the set of the vertexes of the original hypergraph $\mathcal{H}$, and the union of edges on each node of T is equal to the set of the edges of $\mathcal{H}$. In other words, the decomposition method *det-k-CP* does not lose any constraint or variable.

Having these two conditions we can easily validate whether a given decomposition tree is successfully decomposed by *det-k-CP*. Besides, if a decomposition tree satisfies the conditions for *det-k-CP*, it must meet the four conditions defined in [8] for hypertree decomposition (see Sect. 2).

Proposition 1. *A degenerate decomposition tree T of a hypergraph H generated by det-k-CP is a hypertree decomposition of the hypergraph H.*

Proof of Proposition 1. To prove this proposition, we are going to check and confirm the four conditions of *hypertree decomposition* one by one.

---

[2] Please note that the verb "eliminate" does not mean an edge is deleted, it means that we can ignore the join selection for the nodes connected by this edge.

i Since *det-k-CP* does not remove any constraint, as mentioned in the second condition of *det-k-CP*, for every constraint e in the hypergraph H, we can find a node p in the degenerate decomposition tree T, where $\chi(p)$ contains e.

ii The second condition of *hypertree decomposition* ensures that all nodes that share a common vertex v of H induce a connected subtree of T. To prove it by contradiction, we assume there is a vertex v in T, where all nodes that contain v cannot induce a subtree of T. Therefore, in that case, these nodes result in a circuit which indicates there exist edges induced by a set of vertexes that cannot be eliminated by an alternative edge. This is contradicted by the condition 1 of *det-k-CP* in which the shared variables between any pair of adjacent nodes must contain the shared variables between non-adjacent nodes.

iii Because *det-k-CP* does not remove any variable (vertex), for any node p in T, $\chi(p) = (vertexes)(\cup\lambda(p))$, which satisfies $\chi(p) \subseteq (vertexes)(\cup\lambda(p))$.

iv For a given node p in T, $(vertexes)(\cup\lambda(p)) \cap \chi(T_p) = (vertexes)(\cup\lambda(p)) = \chi(p)$, which satisfies the fourth condition of *hypertree decomposition*. $\square$

Algorithm 1. We are now going to explain the *det-k-CP* in more detail. In order to relatively evenly distribute workloads, line 1 of Algorithm 1 sorts the constraints based on its computational requirements (weight), which depend on many factors, such as, the time complexity of constraint propagator used by the constraint solver, the number of variables of the constraint, and the range of each of these variables. Then, after sorting procedure the constraints are inserted into an array (*array_Nodes*) with length k in turn in line 5 so that each node of the array contains the same amount of workload. Algorithm 1 runs into the loop in line 7–13 until a qualified solution is found. Since sometimes a qualified solution cannot be obtained in one iteration, we use the heuristic for swap procedure in line 12 of Algorithm 1. The heuristic has many choices, for example, random exchange, switching two nodes that have the fewest and the most number of constraints, or exchanging nodes based on the permutation in lexicographic order of the indexes of *array_Nodes* in turn. For instance, if the length of *array_Nodes* is 4, we might first use the permutation (0, 1, 2, 3), then (0, 1, 3, 2) and so on.

Algorithm 2. Let us now consider the function *getPotentialSolution* defined by Algorithm 2, which exhaustively invokes Algorithm 3 for all the edges between non-adjacent nodes. To this aim, the starting point of the non-adjacent edge (*i_start*) is assigned the third to last index in line 2 of Algorithm 2, which should be the first unconnected node with the last node (with index $i\_len - 1$). Then, it is decremented on each iteration until it reaches the first index of the array in the outer loop; and the ending point of the non-adjacent edge is initialized as $i\_end = i\_start + 2$, then the inner loop continues to iterate to the end of the the array ($i\_len - 1$).

Algorithm 1. *det-k-CP(N,k)*

Input: A Constraint Network N, and the desired number of nodes k.
Output: A degenerate hypertree with k nodes
1 Set *list_LN* = a list which contains sorted constraints of N based on weight ;
2 Set *i_size* = the size of *list_LN* ;
3 Initialize an array *array_Nodes* with k nodes ;
4 **for** $i \leftarrow 1$ *to* *i_size* **do**
5 | add *list_LN*[i] into *array_Nodes*[$i\%k$];
6 **end**
7 **while** *true* **do**
8 | getPotentialSolution(*array_Nodes*);
9 | **if** *array_Nodes pass test conditions (1) and (2)* **then**
10 | | **break**;
11 | **end**
12 | Swap nodes in *array_Nodes*;
13 **end**
14 **return** *array_Nodes*;

Algorithm 2. *getPotentialSolution(array_Nodes)*

Input: *array_Nodes*
Output: A potential solution
1 Set *i_len* = the length of *array_Nodes* ;
2 Set *i_start* = *i_len*-3;
3 **while** *i_start* ≥ 0 **do**
4 | **for** *i_end* $\leftarrow$ *i_start* $+ 2$ *to* *i_len* $- 1$ **do**
5 | | eliminateEdge(*i_start*, *i_end*, *array_Nodes*) ;
6 | **end**
7 | Set *i_start* = *i_start* $- 1$;
8 **end**

Algorithm 3. The function *eliminateEdge* plays an important role in *det-k-CP*. For two non-adjacent nodes N_{i_start} and N_{i_end}, Algorithm 3 might add constraints to the nodes between N_{i_start} and N_{i_end} so that a potential edge between N_{i_start} and N_{i_end} could be covered. In line 2 of Algorithm 3, *list_2Eliminated* is set to all shared variables between non-adjacent nodes N_{i_start} and N_{i_end}. For each edge, which is caused by shared variables between N_{i_start} and N_{i_end}, *eliminateEdge* checks whether or not every edge between adjacent nodes N_i and N_j, where i is in the interval $\{i_{start}, ..., i_{end} - 1\}$ and $j = i + 1$, contains all variables that are also included in the input edge between nodes N_{i_start} and N_{i_end}, as shown in line 5 of Algorithm 3.

If an edge between N_i and N_j contains all shared variables that the edge between N_{i_start} and N_{i_end} has (in line 5), then the *for* loop runs into the next iteration for the next edge between N_{i+1} and N_{j+1}; otherwise, line 6 removes all shared variables of N_i and N_j on *list_2Eliminated*.

Algorithm 3. *eliminateEdge(i_start, i_end, array_Nodes)*

Input: *i_start, i_end, array_Nodes*

1 **for** $i \leftarrow i\_start$ **to** $i\_end - 1$ **do**
2   Set *list_2Eliminated* = getSharedVariables(*i_start, i_end, array_Nodes*);
3   Set $j = i + 1$;
4   Set *list_SharedOnMainPath* = getSharedVariables(*i, j, array_Nodes*);
5   **if** *list_SharedOnMainPath.notContainsAll(list_2Eliminated)* **then**
6   *list_2Eliminated.removeAll(list_SharedOnMainPath)*;
7   Set *list_2beAddedConstraints* = *getMinimumSetConstraints4Share(i, array_Nodes, list_2Eliminated)*;
8   **foreach** *constraint cs* $\in$ *list_2beAddedConstraints* **do**
9   **if** *array_Nodes[j]* *notContained cs* **then**
10   add *cs* into *array_Nodes[j]*;
11   **end**
12   **end**
13   **end**
14 **end**

The function *getMinimumSetConstraints4Share*, which will be presented later in Algorithm 4, returns the minimum number of constraints that covers all variables on the list *list_2Eliminated*. In line 8–12, we loop through all constraints obtained by *getMinimumSetConstraints4Share*, if the constraint is not contained in N_j, the constraint will be added into N_j. By doing so, the edges between non-adjacent nodes N_{i_start} and N_{i_end} can be eliminated. Because the shared variables between N_{i_start} and N_{i_end} are now covered by all edges between adjacent nodes N_i and N_j with $\{i_{start}, ..., i_{end} - 1\}$ and $j = i + 1$.

Algorithm 4. The idea behind this function, shown in line 4–16 of Algorithm 4, is that it, in each iteration for each variable in the difference of set (*list_2Eliminated*) and (*list_SharedOnMainPath*), searches the constraints that covers as many variables as possible on (*list_2Eliminated*); meanwhile, the constraints themselves contain as few variables as possible. In line 12 of Algorithm 4, the covered variables on (*list_2Eliminated*) are added into the HashSet variable *hashSet_IsCovered*, which helps avoid rechecking covered variables at the beginning of the *for* loop (line 5–7).

So far we have discussed the detailed process of *det-k-CP*. The reason why we introduce randomness into the search process to *Swap* method in line 12 of Algorithm 1 is that a large number of loops would be incurred if we backtracked to edges eliminated before but the newly added constraints make the edges appear again. For instance, in Fig. 3, we added some constraints onto node 4 due to the elimination process for the edge between node 3 and 5, then the edge between node 4 and 7, which had been removed before, happened to occur again, and this edge forced us to add new constraints onto node 5 and 6, after

---

Algorithm 4. *getMinimumSetConstraints4Share(i, array_Nodes, list_2Eliminated)*

Input: *endPoint, array_Nodes, list_2Eliminated*

Output: A set of constraints that has minimal number of variables to cover the
   list_2Eliminated

1   Set *list_SharedConstraints* = add all constraints which covers the variables in
   the *list_2Eliminated* and remove duplicates;

2   Initialize *list_Result* ;

3   Initialize *HashSet HashSet_IsCovred* ;

4   **foreach** *Variables v ∈ list_2Eliminated* **do**

5 | **if** *HashSet_IsCovred contains v* **then**

6 | | continue;

7 | **end**

8 | **foreach** *Constraint c ∈ list_SharedConstraints* **do**

9 | | add the constraint c' into *list_Result*, which covers the maximum
   number of variables in the *list_2Eliminated* and the constraint itself
   contains minimum number of variables;

10 | | **foreach** *Variables $v' ∈ c'$* **do**

11 | | | **if** $v' ∉ HashSet\_IsCovred$ **then**

12 | | | | add v' into *HashSet_IsCovred*;

13 | | | **end**

14 | | **end**

15 | **end**

16   **end**

17   **return** *list_Result*;

---

that the edge between node 3 and 5 occurred again, consequently, we would fall
into repeated elimination process until the worst case happened in which each
node filled up with all the constraints of the given constraint network. Now, we
would like to analyze the time complexity of *det-k-CP*. If we combine the loops
in Algorithm 2 with the loop in Algorithm 3 to form a triple-nested loop, the
total number of iterations for the triple-nested loop can be calculated by the
following recurrence relation:

$$n_3 = 2 \tag{3}$$

$$n_4 = 3 + 2 + n_3 \tag{4}$$

$$n_5 = 4 + 3 + 2 + n_4 \tag{5}$$

$$\vdots$$

$$n_k = (k-1) + (k-2) + \cdots + 2 + n_{k-1} = \frac{(k^2 - k - 2)}{2} + n_{k-1} \tag{6}$$

Where n_k denotes the number of iterations for the triple-nested loop for k
number of nodes of the target decomposition tree. A recurrence relation for $\{n_k\}$
can be obtained by considering whenever one node is added to a tree which has
$k - 1$ nodes; therefore, new non-adjacent nodes are generated, consequently, we
have to eliminate these edges, where the number of these edges can be summed

by $(k-1)+(k-2)+\cdots+2$. For instance, as can be seen in Fig. 4, when node 0 is added to the original tree, the number of times of elimination process is increased by 5. To obtain the explicit formula for this recurrence relation, we solve it with the initial conditions $n_3 = 2$, $n_2 = 0$ and $n_1 = 0$. The solution of the recurrence relation is $n_k = \frac{k^3-7k+6}{6}$, which means the number of loops of the triple-nested loop is exactly $\frac{k^3-7k+6}{6}$.

At each iteration of the triple-nested loop, from line 2 to 13 of Algorithm 3, the number of executions can be bounded by the number of constraints (N_c) plus the complexity of method $getMinimumSetConstraints4Share$, denoted by $\mathcal{O}(N_c) + \mathcal{O}(getMinimumSetConstraints4Share)$. The implementation of $getMinimumSetConstraints4Share$ is bounded by $\mathcal{O}(N_v \cdot N_c)$, where N_v is the number of variables. Thus, Algorithm 2 is bounded by $\mathcal{O}(\frac{k^3-7k+6}{6} \cdot (N_v \cdot N_c + N_c))$. Algorithm 1, which can be viewed as the outermost loop of the entire algorithm, can be bounded by $\mathcal{O}(k \cdot (N_c - \frac{N_c}{k}))$. This is because the loop of Algorithm 1 eventually stops running when each node is added to contain the whole constraint network. The overall time complexity is, therefore, $\mathcal{O}(k \cdot (N_c - \frac{N_c}{k}) \cdot \frac{k^3-7k+6}{6} \cdot (N_v \cdot N_c + N_c))$. Therefore, the asymptotic time complexity is $\mathcal{O}(k^4 \cdot N_c^2 \cdot N_v)$. Note that the runtime may be significantly smaller in practice since we take into account the worst cases for Algorithms 1 and 3.

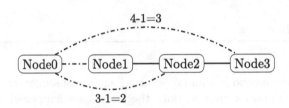

Fig. 4. A new node is prepended to a degenerated tree with 3 nodes.

4   Experimental Results

In this section, we present our experimental results of the algorithm $det\text{-}k\text{-}CP$ when applied to the benchmark suite provided by Gottlob et al. used in [1]. Note that we do not compare $det\text{-}k\text{-}CP$ with $det\text{-}k\text{-}decomp$ because these two algorithms aim at two different decomposition targets as mentioned before. However, this should not place an obstacle for us since the hypergraphs in the benchmarks are extracted from practical industrial constraint satisfaction problems. All the experiments are set up on an iMac computer having an Intel i7-3770 CPU, 3.40 GHz, with 8 GB 1600 MHz DDR3 and running under macOS Sierra version 10.12.5. The algorithms are implemented in Java under JDK version 1.8.0_131.

Since the running platform of hardware for $det\text{-}k\text{-}CP$ will be multi-core processors, we choose 4, 8, and 16 for k as the number of nodes for the target degenerated decomposition tree. Because all the hypergraphs in the benchmark suite are decomposable by *det-k-CP*, we only pay attention to the experiment data that includes the execution time, the number of constraints on each node and the variance of the number of constraints among the different nodes, and as well as the impact of different stochastic strategies on the experiment results. At current stage, we regard that the variance of the number of constraints is an important target for parallel constraint solving. However, it will be easy to switch to the variance of the weight measured by the amount of computation of each node if we can estimate the amount of computation of a constraint in a given constraint network in the future.

Table 1 shows the experimental results for *det-k-CP* of the benchmark suite from [1]. The symbols ●, ▲, and ■ denote the benchmark packages Daimler-Chrysler, Grid2D, and ISCAS89 as used in [1], respectively. The number of vertexes and edges of each original instance is left out here due to lack of space. For the details on the benchmarks, we refer to [1], Sect. 5.

In Table 1, T represents execution time in *ms* (unless other specified), and Min, Max stands for the minimal and maximal number of constraints among all nodes, respectively. The variance of the number of constraints of all nodes is denoted σ^2. There are three different background colors that indicate three types of heuristics for the swap procedures for each instance, which are a permutation[3], random exchange, and switching nodes according to the number of constraints in turn.

Though Table 1 omits some instances from the benchmark suite due to lack of space, all the instances are decomposable by *det-k-CP*. Besides, the execution time of instances is affordable even for the largest instance *s5378*, which has 2958 constraints and 2993 variables. Nevertheless, for large-scale instances, such as *s5378, grid2d_75* etc. the algorithm *det-k-CP* does not achieve one principal goal of decomposition that each node has relativity balanced workload distribution. The maximal node contains 1220 constraints in overall 2958 constraints, which is probably impossible to be solved by any single thread constraint solver for a practical problem based on our user experience of constraint solver.

We do not include small instances such as *adder_15, adder_25* etc. from the benchmark suite because it turns out that these small constraint networks are solvable for mainstream constraint solver and do not require parallel solving.

The decomposition results for the medium scale instances, such as *adder_75, bridge _75, NewSystem3, NewSystem4,* and *S953* with a number of constraints ranging from around 400 to 700, might be suitable for parallel constraint solving. Because each node for these instances can be solved within a reasonable time since the maximum number of constraints of these instances is slightly greater than 200 such as 201 in *adder_75* and 210 *NewSystem4*, and the total number of constraints for *adder_75* and *NewSystem4* are 677 and 418, respectively. In addition to the relatively small variance of these instance indicates the number of constraints for these instances is not spread out from their mean. As we mentioned before, whether or not a given constraint network can be solved within

---

[3] The term *permutation* is explained in Algorithm 1.

Table 1. Experimental results for *det-k-CP* of the benchmark suite from [1].

Instance	$k = 4$				$k = 8$				$k = 16$			
	T	Min	Max	σ^2	T	Min	Max	σ^2	T	Min	Max	σ^2
●adder_75	75	94	201	1501	309	47	180	1832	3.7s	24	162	1874
●adder_75	9	94	173	1053	134	47	160	1734	3.9s	23	162	1881
●adder_75	10	94	208	1815	192	47	207	3201	3.6s	24	217	3608
●adder_99	75	124	262	2341	406	62	236	3062	5s	31	207	3050
●adder_99	10	124	261	4430	242	62	204	2098	12s	31	303	7291
●adder_99	14	124	272	3050	263	62	264	5008	5.9s	31	84	5724
●bridge_50	59	113	211	1323	277	57	209	2517	3s	29	185	2032
●bridge_50	7	113	221	2760	165	56	207	3160	3s	28	149	2633
●bridge_50	9	113	227	1830	207	57	219	2841	3.3s	29	209	3017
●bridge_75	82	170	314	2934	511	85	294	4968	7.8s	43	267	4938
●bridge_75	14	169	303	2729	639	84	347	7046	9s	42	322	5869
●bridge_75	17	170	336	4065	402	85	321	6796	8s	43	332	8396
●bridge_99	172	224	420	5168	13	112	376	7484	16.7s	56	358	8983
●bridge_99	38	223	435	6166	782	111	430	16265	14s	55	387	13914
●bridge_99	33	224	442	6727	737	112	424	11258	17s	56	428	12717
●NewSystem2	54	50	102	341	159	25	88	387	1.13	53	86	450
●NewSystem2	5	50	91	249	81	25	116	723	1.5	12	110	1000
●NewSystem2	3	50	102	341	72	25	111	729	979	13	97	636
●NewSystem3	84	70	133	343	268	35	134	1046	15s	18	157	1709
●NewSystem3	5	69	145	1370	150	34	141	1824	3.9s	12	159	2242
●NewSystem3	7	70	149	899	115	35	141	1240	2m	18	146	1681
●NewSystem4	71	105	211	1467	391	53	185	1610	4.5s	27	166	1717
●NewSystem4	11	104	210	1629	328	52	235	5516	75s	26	264	3585
●NewSystem4	10	105	211	1467	286	53	219	3102	5s	27	228	3778
▲grid2D_40	97	200	332	2630	777	100	311	4949	13S	50	300	6002
▲grid2D_40	28	200	335	2755	695	100	320	8010	15S	50	302	6269
▲grid2D_40	31	200	338	3079	592	100	318	5297	14S	50	337	7957
▲grid2D_75	486	703	1149	29669	7.8s	352	1093	62397	3.7m	176	1057	77453
▲grid2D_75	277	703	1192	35509	9.4s	351	1149	66461	3.7m	175	1078	59942
▲grid2D_75	298	703	1186	35874	7.9s	352	1131	68936	3.5m	176	1154	91808
■s953	71	106	204	1299	313	53	195	2286	3s	27	172	1965
■s953	8	106	211	2756	225	53	217	2776	3.2s	26	194	3226
■s953	6.9	106	204	1298	170	53	205	2724	3.5s	27	218	3638
■s1494	78	164	335	3859	523	82	291	4489	6.6s	41	249	3938
■s1494	15	163	350	8079	380	81	345	11411	7.6s	40	348	13751
■s1494	17.5	164	348	4832	340	80	332	7630	7.6s	41	339	9316
■s5378	507	740	1461	70147	5.8s	370	1338	101431	3m	185	1220	103192
■s5378	209	739	1347	57433	13.2s	369	1642	153866	5.3m	184	1568	167284
■s5378	183	740	1461	70148	6.8s	370	1614	176602	3.8m	185	1565	178186

a reasonable time does not only depend on the number of constraints, but also on the time complexity of propagators employed by the constraint solver, the number of variables of the constraint network and the size of the range of the variables.

The results in Table 1 show that the method of heuristics has a significant impact on the results of decomposition. In most cases, the exchange of nodes by permutation order (first lines, ie. white background color) gets smaller variances, but there are exceptions, such as *bridge_75, NewSystem4*. It should be noted that all instances in the benchmark suite are decomposable by *det-k-CP* is because they can be decomposed by *det-k-decomp*. Recall that in the proof of Proposition 1, we have shown that if a hypertree decomposition meets the two conditions of *det-k-cp*, then it also satisfies the four conditions of *det-k-decomp*. Moreover, the width of hypertree as a result decomposed by *det-k-CP* far outweighs the width hypertree as a result decomposed by *det-k-decomp*. Therefore, if an instance can be decomposed by *det-k-decomp* with small width[4], it must be decomposed by *det-k-CP* with a relatively big width (e.g., 50). As it was mentioned previously, the extreme cases, where a node or more than one node of the decomposition tree saturated with the entire constraint network or the distribution of constraints is relatively unbalanced (e.g.,*grid2d_75*), result in failure because the goal of *det-k-cp* is a relatively even distribution of workload for parallel constraint solving.

In summary, we can conclude that *det-k-CP* can decompose a given constraint network within a reasonable execution time except for very large instance (e.g., *s5378* with 2958 constraints and 2993 variables) and a big k (e.g., $k = 16$). For the application of parallel constraint solving, the algorithm can be applied to medium scale constraint networks with the number of constraints ranging from around 400 to 700.

5   Conclusion and Future Work

We have presented the new algorithm *det-k-CP* to construct a degenerate decomposition tree for parallel constraint solving, and we have also evaluated *det-k-CP* by a benchmark suite from previous research *det-k-decomp*. Our results have shown that it is appropriate for *det-k-CP* to evenly partition a constraint network with around 400 to 700 constraints for a distribution on a given number of parallel constraint solving cores. However, we believe that there is a lot potential to improve *det-k-CP*. For example, the algorithm should take into consideration an estimate of the amount/complexity of computation for the constraints so that we could choose eg., constraints with low computation requirement when adding constraints to another node in Algorithm 3. Furthermore, local search methods such as tabu search can be employed to replace the existing stochastic strategies in Algorithm 1 in order to obtain more optimized decomposition result. Besides, for balanced workload distribution, we can add constraints to one node

---

[4] We observed all the instances can be successfully decomposed by *det-k-decomp* when the width is 2.

from other nodes covering the same variables to preserve the decomposition tree. Finally, the key indicator of the value of this research depends on whether we can obtain speedup or even super-linear speedup when using *det-k-CP* for parallel constraint solving, which is to be researched in detail.

Acknowledgments. We would like to express our special thanks to Georg Gottlob and Wolfgang Fischl for their source code of *det-k-decomp*, especially the benchmark suite for hypertree decomposition.

References

1. Gottlob, G., Samer, M.: A backtracking-based algorithm for hypertree decomposition. J. Exp. Algorithmics **13**, 1 (2009)
2. Gottlob, G., Leone, N., Scarcello, F.: Advanced parallel algorithms for acyclic conjunctive queries. Technical Report DBAI-TR-98/18 (1998). http://www.dbai.tuwien.ac.at/staff/gottlob/parallel.ps
3. Gottlob, G., Leone, N., Scarcello, F.: The complexity of acyclic conjunctive queries. J. ACM (JACM) **48**(3), 431–498 (2001)
4. Gottlob, G., Greco, G., Leone, N., Scarcello, F.: Hypertree decompositions: questions and answers. In: Proceedings of the 35th ACM SIGMOD-SIGACT-SIGAI Symposium on Principles of Database Systems, pp. 57–74. ACM (2016)
5. Dechter, R.: Constraint Processing. Morgan Kaufmann, Burlington (2003)
6. Dechter, R., Pearl, J.: The cycle-cutset method for improving search performance in AI applications. In: Third IEEE Conference on AI Applications, pp. 224–230. IEEE (1987)
7. Gyssens, M., Paredaens, J.: A decomposition methodology for cyclic databases. In: Gallaire, H., Nicolas, J., Minker, J. (eds.) Advances in Data Base Theory, vol. 2, pp. 85–122. Plemum Press, New York (1984). https://doi.org/10.1007/978-1-4615-9385-0_4
8. Gottlob, G., Leone, N., Scarcello, F.: Hypertree decompositions and tractable queries. In: Proceedings of the Eighteenth ACM SIGMOD-SIGACT-SIGART Symposium on Principles of Database Systems, pp. 21–32. ACM (1999)
9. Gottlob, G., Leone, N., Scarcello, F.: A comparison of structural CSP decomposition methods. Artif. Intell. **124**(2), 243–282 (2000)
10. Prud'homme, C., Fages, J.G., Lorca, X.: Choco Documentation. TASC, INRIA Rennes, LINA CNRS UMR 6241, COSLING S.A.S. (2016)
11. Gottlob, G., Leone, N., Scarcello, F.: Robbers, marshals, and guards: game theoretic and logical characterizations of hypertree width. J. Comput. Syst. Sci. **66**(4), 775–808 (2003)

Declarative Systems

Declarative Aspects in Explicative Data Mining for Computational Sensemaking

Martin Atzmueller$^{(\boxtimes)}$

Department of Cognitive Science and Artificial Intelligence, Tilburg University,
Warandelaan 2, 5037 AB Tilburg, Netherlands
m.atzmuller@uvt.nl

Abstract. Computational sensemaking aims to develop methods and systems to "make sense" of complex data and information. The ultimate goal is then to provide insights and enhance understanding for supporting subsequent intelligent actions. Understandability and interpretability are key elements of that process as well as models and patterns captured therein. Here, *declarativity* helps to include guiding knowledge structures into the process, while *explication* provides interpretability, transparency, and explainability. This paper provides an overview of the key points and important developments in these areas, and outlines future potential and challenges.

Keywords: Computational sensemaking · Data mining
Declarative modeling · Domain knowledge · Explicative data analysis
Knowledge graph · Statistical relational learning

1 Introduction

Computational sensemaking aims to "make sense" in the context of complex information and knowledge processes. This is enabled using computational methods for *analysis, interpretation,* and *intelligent decision-support.* While the latter is mostly supported by human-computer interaction techniques, the former two are supported by data mining approaches, in particular, *explicative data mining* methods.

Overall, data mining systems are commonly applied to obtain a set of *novel, potentially useful,* and ultimately *interesting* patterns from (large) data sets [27]. While the resulting patterns are typically interpretable, e.g., in pattern mining, the large result sets of potentially interesting patterns that the user needs to assess, require further *exploration* and *interpretation* techniques. In general, facilitating the understandability and interpretability of the process as well as its "products" (e.g., in the form of patterns) need to consider two important aspects: *declarativity*, in order to include guiding knowledge structures into the process, as well as *explication* in order to provide interpretability, transparency, and explainability. Both *declarative* as well as *explicative* approaches work together in that

© Springer Nature Switzerland AG 2018
D. Seipel et al. (Eds.): DECLARE 2017, LNAI 10997, pp. 97–114, 2018.
https://doi.org/10.1007/978-3-030-00801-7_7

context, complementing each other. This paper provides an overview of the key points and important developments in these areas, and outlines future potential and challenges.

2   Declarative Aspects in Explicative Data Mining

While declarative approaches allow for the incorporation of background knowledge and the guidance of the data mining process, explicative data mining [3] focuses specifically on obtaining interpretable models and patterns, on transparency on the data mining process, and on explanainable or explanation-aware mining. In that way, both complement each other quite well, such that declarative aspects can be incorporated into explicative data mining for enhancing interpretability, transparency, and explainability.

Below, we briefly introduce declarative aspects on data mining, especially focussing on the modeling of background knowledge and the specification of knowledge to be incorporated into the data mining process. For that, we first introduce explicative data mining methods, including exploratory and explanation-aware approaches. Here, we discuss examples in the context of pattern mining methods [1, 2, 4, 13, 45, 94], since pattern mining is a prominent research direction for obtaining interpretable patterns, enabling a transparent data mining process. In particular, we discuss the relation to incorporating prior knowledge, e.g., in the form of knowledge patterns [12] and knowledge graphs [16, 36, 92] into the data mining process. This enables hybrid approaches that incorporate semantic knowledge into the process, e.g., supporting modeling and explanation methods.

2.1   Explicative Data Mining

Data mining methods are commonly applied to obtain a set of *novel, potentially useful*, and ultimately *interesting* patterns from (large) data sets cf., [27]. This can be achieved e.g., utilizing exploratory data mining techniques like association rule mining or subgroup discovery, as sketched above.

However, most common data mining methods and approaches lack important aspects, i.e., *interpretability, transparency* and *explainability* in order to be *explicative* towards its users. Especially considering complicated black-box models this becomes relevant, e.g., when providing recommendations and filtering. Prominent application examples include, for example, large online social networks, e.g., when providing posts or news to users, but also in predictive settings such as user scoring or classification in e-commerce. Here, intransparent methods and models make it more difficult to spot mistakes and can lead to biased decisions, e.g., based on incorrect training data; in general, they stretch the trust humans have (and should rightfully have) in the respective predictions. Then, the potential competitive advantage through better predictions for humans, for businesses, and for society as a whole comes at the cost of reduced explanatory power.

This is particularly important in the light of the European Union's new General Data Protection Regulation, which will as of this year enforce a "right to explanation" (providing users the right to obtain an explanation for any algorithmic decisions that were made about them). Overall, there will be a major impact on business, technology, and society. In particular, in the area of data mining, these developments give rise to major research challenges and a major impact on the interaction of humans with such algorithms and according technology in itself, cf., [31].

Explicative data mining [3] is a comprehensive paradigm for interpretable, transparent and explainable data analysis. Similar to the philosophical process of *explication* cf., [19,54] which aims to make the implicit explicit, explicative data mining aims to model, describe and explain the underlying structure in the data.

Explicative data mining targets interpretable (and transparent) models utilizing exploratory and explanation-aware methods. These can be constructed and inspected on different layers and levels. This ranges from pure data summarization to pattern-based exploratory data mining. Furthermore, these features also provide for different options for including the human in the loop, e.g., using visualization methods embedded into interactive and semi-automatic approaches and methods.

Below, we discuss how to include declarative aspects into explicative data mining, focussing on interpretable models as well as explainable or explanation-aware approaches. For the former, we focus on how to model and provide explication knowledge that is then integrated into the data mining process. We take a pragmatic view, and consider the typical data mining process, e.g., structured according to the CRISP-DM [20,93] cycle, "as is" – and thus incorporate important elements of purely declarative approaches, e.g., [17,18]. We first focus on exploratory approaches, before we discuss explanation-aware methods.

2.2 Exploratory Data Mining

In the scope of explication, exploratory data mining techniques can provide a first view on the data in order to detect interesting patterns. Exploratory techniques span from statistical approaches for characterizing a dataset or determining key (influence) factors, e.g., [29,89] to more refined (semi-)automatic approaches, e.g., for *local pattern detection* [43,58,59]. Local pattern detection aims to discover *local models* characterizing (or describing) parts of the data given an interestingness measure, e.g., [43]. In addition, interactive visualization methods, e.g., [28,30,39,42,76,78,79,82,86] can be applied (or combined with automatic methods) for further supporting data exploration.

Pattern Mining. Overall, a broadly applicable and powerful set of methods is provided by the area of pattern mining. Common methods include those for association rule mining [1] or subgroup discovery, e.g., [2,40,94]. The latter is at the intersection of descriptive and predictive data mining [45] and can be applied for a variety of different analytical tasks. Essentially, *subgroup discovery* [2,40,45,94] is an exploratory approach for discovering interesting subgroups – as an instance of local pattern detection [2,43,48,58,59]. The interestingness is usually defined by a certain property of interesting formalized by a quality function. Essentially, subgroup discovery is a flexible method for detecting relations between dependent (characterizing) variables and a dependent target concept, e.g., comparing the share or the mean of a nominal/numeric target variable in the subgroup vs. the share or mean in the total population, respectively. The interestingness of a pattern can then be flexibly defined, e.g., by a significant deviation from a model that is derived from the total population. In the simplest case, (see the example above) a binary target variable is considered, where the share in a subgroup can be compared to the share in the dataset in order to detect (exceptional) deviations.

More complex target concepts consider sets of target variables. Here, *exceptional model mining* [2,25,47] focuses on more complex quality functions, considering complex *target models*, like comparing regression models or graph structures. Essentially, exceptional model mining tries to identify interesting patterns with respect to a local model derived from a set of attributes, cf., [23,24]. This can be extended, e.g., for network analysis and (exceptional) graph mining, e.g., [38]. Below, we introduce and define subgroup models (and patterns), as well as association rules more formally.

Domain Knowledge for Semantic Data Mining. In many domains, a lot of (semantic) domain knowledge is available in order to support reasoning processes. However, in data mining, semantic knowledge is scarcely exploited so far. Domain knowledge is a natural resource for knowledge-intensive data mining methods, e.g., [37,70], and can be exploited for improving the quality of the data mining results significantly. Appropriate domain knowledge can increase the representational expressiveness and also focus the algorithm on the relevant patterns. Furthermore, for increasing the efficiency of the search method, the search space can often be constrained, e.g., [9].

There are several approaches, which show how to effectively provide and include domain knowledge into data mining approaches, e.g., [9,22,56,57,60, 64,88] thus supporting explicative data mining by providing semantic specifications. For modeling expected relations for pattern discovery, for example, according methods are presented in [7] utilizing Bayesian network formalizations. For relational data analysis approach, also the comparison of hypotheses with a semantic data model using Bayesian techniques (first order Markov chains) has been targeted in [11]. Furthermore, statistical relational learning, e.g., [21,65,71] combines both probabilistic and complex relational learning approaches, in particular also enabling complex logic-based methods. Such methods then enable powerful declarative approaches in order to provide domain knowledge for data mining.

Easing Knowledge Acquisition Costs. However, knowledge acquisition is often challenging and costly, a fact that is known as the so-called *knowledge acquisition bottleneck*. Thus, an important idea is to ease the knowledge acquisition by reusing existing domain knowledge, i.e., already formalized knowledge that is contained in existing ontologies or knowledge bases. Furthermore, we aim to simplify the knowledge acquisition process itself by providing knowledge concepts that are easy to model and to comprehend.

We propose high-level knowledge, such as properties of ontological objects for deriving simpler constraint knowledge that can be directly included in the data mining step, as discussed, e.g., in [8,9]. Modeling such higher-level ontological knowledge, i.e., properties and relations between domain concepts, is often easier for the domain specialist, since it often corresponds to the mental model of the concrete domain. Below, we outline simple specifications of domain knowledge, before we discuss more complex modeling approaches including integrated knowledge graphs and statistical relational learning approaches.

Subgroups and Association Rules. Subgroup models [9,41], often provided by conjunctive rules, describe'interesting' subgroups of cases, e.g., "the subgroup of 16–20 year old men that own a sports car are more likely to pay high insurance rates than the people in the reference population." Subgroup discovery [2,40,46, 94] is a powerful method, e.g., for (data) exploration and descriptive induction, i.e., to obtain an overview of the relations between a so-called target concept and a set of explaining features. These features are represented by attribute/value assignments, i.e., they correspond to binary features such as items known from association rule mining [1]. As discussed below, in its simplest case the target concept is often represented by a binary variable, but can also extend to more complex target concepts, e.g., considering sets of variables, and their relations.

Formally, a *database* $DB = (I, A)$ is given by a set of individuals I and a set of attributes A. For each attribute $a \in A$, a range $dom(a)$ of values is defined. An attribute/value assignment $a = v$, where $a \in A, v \in dom(a)$, is called a *feature*. We define the feature space V to be the (universal) set of all features.

Basic elements used in subgroup discovery are patterns and subgroups. Intuitively, a *pattern* describes a *subgroup*, i.e., the subgroup consists of instances that are covered by the respective pattern. It is easy to see, that a pattern describes a fixed set of instances (subgroup), while a subgroup can also be described by different patterns, if there are different options for covering the subgroup' instances. In the following, we define these concepts more formally, following an adapted notation of [12].

Definition 1. *A (subgroup) pattern P is defined as a conjunction*

$$P = s_1 \wedge s_2 \wedge \cdots \wedge s_n$$

of (extended) features $s_i \subseteq V$, which are then called selection expressions, where each s_i selects a subset of the range $dom(a)$ of an attribute $a \in A$.

A selection expression s is thus a Boolean function $I \rightarrow \{0, 1\}$ that is true if the value of the corresponding attribute is contained in the respective subset of V for the respective individual.

Definition 2. *A subgroup (extension)* $I_P := ext(P) := \{i \in I | P(i) = true\}$ *is the set of all individuals which are covered by the pattern* P.

The subgroup mentioned above, for example, is described by the relation between the independent (explaining) variables (Sex = male, Age $\leq$ 20, Car = sports car). Furthermore, there is a dependent (target) variable, i.e., Insurance Rate = high; this target variable relates to the *concept of interest* used in subgroup discovery, which is utilized to estimate the interestingness of a subgroup using a quality measure. This is captured by the notion of a subgroup model described below.

In general, the applied quality measure can also be defined using a set of target variables, or more complex models such as Bayesian networks or topological graph structures which relates to the area of *exceptional model mining* [2,25]. In the scope of this paper and our simple example, we focus on simple *binary* target variables given by (simple) features as defined above.

An *association rule*, e.g., [1], is given by a rule of the form $P_B \rightarrow P_H$, where P_B and P_H are patterns; the rule body P_B and the rule head P_H specify sets of items. For the insurance domain, for example, we can consider an association rule showing a combination of potential risk factors for high insurance rates and accidents:

$$\text{Sex} = \text{male} \wedge \text{Age} \leq 20 \wedge \text{Car} = \text{sports car}$$
$$\rightarrow \text{Insurance Rate} = \text{high} \wedge \text{Accident Rate} = \text{high}$$

A *subgroup model* is a special association rule, namely a horn clause $P \rightarrow e$, where P is a pattern and the feature $e \in V$ is called the target variable. For subgroup discovery, a fixed rule head is considered.

In general, the quality of an association rule is measured by its support and confidence, and the data mining process searches for association rules with arbitrary rule heads and bodies, e.g., using the apriori algorithm [1]. For subgroup models there exist various (more refined) quality measures, e.g., [2,40]: Since an arbitrary quality function can be applied, the anti-monotony property of support used in association rule mining cannot be utilized in the general case.

The applied quality function can also combine the difference of the confidence and the apriori probability of the rule head with the size of the subgroup. Since mining for interesting subgroup patterns is more complicated, usually a fixed, atomic rule head is given as input to the knowledge discovery process.

Declarative Specifications of Domain Knowledge. As we have presented in [12], a prerequisite for the successful application and exploitation of domain knowledge is given by a concise *declarative* specification of the domain knowledge. A concise specification also provides for better documentation,

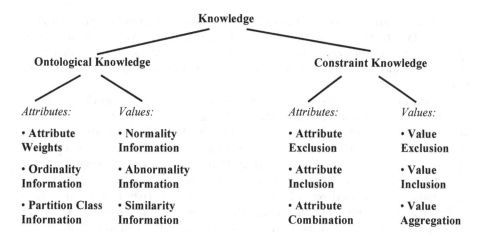

Fig. 1. Hierarchy of (abstract) knowledge classes and specific types, cf., [12].

extendability, and standardization. Below, we summarize the approaches proposed in [12] and provide examples of its instantiation in the field of pattern mining as outlined above.

In contrast to existing approaches, e.g., [70, 95] we focus on domain knowledge that can be easily declared in symbolic form. Furthermore, the presented approach features the ability of deriving *simpler* low-level knowledge (constraints) from high-level ontological knowledge. In general, the search space considered by the data mining methods can be significantly reduced by *shrinking* the value ranges of the attributes. Furthermore, the search can often be focused if only *meaningful* values are taken into account. This usually depends on the considered ontological domain.

The considered classes of domain knowledge include ontological knowledge and (derived) constraint knowledge, as a subset of the domain knowledge described in [8, 9]. Figure 1 shows the knowledge hierarchy proposed in [12], from the two knowledge classes to the specific types, and the objects they apply to.

Prolog-Based Specifications. For specifying the properties and relations of the concepts contained in the domain ontology, we utilize Prolog rules as a compact and versatile representation, cf., [12]. Using these rules, we obtain a suitable representation formalism for ontological knowledge. Using these, we can automatically derive ad-hoc relations between ontological concepts using (simple) rules.

Essentially, the declarative features of Prolog allow simple and transparent knowledge specification, integration and advancement: Depending on the experience of the domain specialist, new knowledge can be added extending the existing knowledge, new relations can be introduced, and furthermore additional advanced features like derivation rules can be directly implemented using Prolog. In addition, using domain specific languages built on top, e.g., [75, 85] the declarativity can even be further enhanced, while also providing an even simpler interface to the domain specialist. Then, this both provides for a concise specification and also comprehensive overview, documentation and summary for the domain specialist, which is typically easy to comprehend, to interpret and to extend.

Below, we focus on selected examples proposed in [12]. For brevity, we focus on simple examples considering attributes, e.g., using attribute weights, and attribute inclusion/exclusion constraints, cf., Fig. 1. With these, attributes can be selected (or excluded) such that they do not occur in patterns constructed by the applied pattern mining method. In that way, for example, exclusion constraints restrict the pattern space. Also, combination constraints inhibit the examination of specified sets of concepts. In that way, they help to find more understandable results. For increasing the representational expressiveness and thus the interpretability of patterns, modifications of the considered attributes (and their combinations) can be utilized to make the discovered patterns and models more meaningful for the user. It is easy to see, that the specifications regarding combinations of attributes, and their explicit exclusion and inclusion directly map to attribute values, and features, respectively. Then, regarding the presented pattern mining methods the format of the considered patterns, and their "building blocks" can be conveniently.

Examples. As outlined above, we summarize some simple examples regarding the declarative specifications presented in [12], for which we refer to for an in-depth description and detailed discussion.

As a first example, *partition class information*, provides semantically distinct groups of attributes. These disjoint subsets usually correspond to certain problem areas of the application domain. E.g., in the medical domain such partitions are representing different organ systems like *liver, kidney, pancreas, stomach, stomach,* and *intestine.* For each organ system a list of attributes is given:

```
attribute_partition(inner_organs, [
    [fatty_liver, liver_cirrhosis, ...],
    [renal_failure, nephritis, ...], ... ]).
```

Furthermore, *attribute weights* denote the relative importance of attributes, and are a common extension for knowledge-based systems [14]. In the car insurance domain, for example, we can state that the attribute *Age* is more important than the attribute *Car Color*, since its assigned weight is higher:

```
weight(age, 4).
weight(car_color, 1).
```

Deriving Constraints. We can construct attribute exclusion constraints using attribute weights to filter the set of relevant attributes by a weight threshold or by subsets of the weight space.

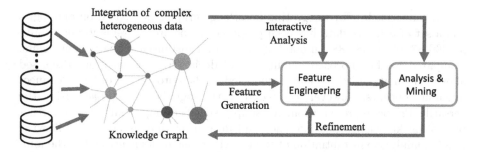

Fig. 2. Overview on a framework for mixed-initiative feature engineering and data mining using knowledge graphs, cf., [13] for a detailed discussion.

```
dsdk_constraint(exclude(attribute), A) :-
    weight(A, N), N =< 1.
dsdk_constraint(include(attribute), A) :-
    weight(A, N), N > 1.
```

Partition class information can be used to infer attribute combination constraints in order to prevent the combination of individual attributes that are contained in separate partition classes. Alternatively, inverse constraints can also be derived, e.g., to specifically investigate inter-organ relations in the medical domain.

```
dsdk_constraint(exclude(attribute_pair), [A1, A2]) :-
    attribute_partition(_, P),
    member(As1, P), member(As2, P), As1 \= As2,
    member(A1, As1), member(A2, As2).
```

Finally, we can use a generic Prolog rule for detecting conflicts w.r.t. these rules and the derived knowledge (*automatic verification*):

```
dsdk_constraint(error(include_exclude(X), Y) :-
    dsdk_constraint(include(X), Y),
    dsdk_constraint(exclude(X), Y).
```

Knowledge Graphs. A further effective approach for modeling explication knowledge is given by constructing a knowledge graph cf., e.g., [16,36]: Here, the data is integrated into a comprehensive knowledge structure capturing the relations between concepts and their properties in an explicit way, cf., [16,36,72, 92]. Then, this structure can be exploited in order to facilitate data mining, e.g., by applying ontologies in the data mining step. However, so far the approaches only apply a "shallow" coupling, that is, typically there is no deep integration of knowledge graph and mining approach (Fig. 2).

First approaches for integrating knowledge graphs, i.e., based on ontologies and a set of instance data has been proposed in the area of semantic data mining [66,87,88]. In [87,88] an ontology is used for instantiating pattern elements. Compared to the approach making use of declarative specifications discussed above, this makes of the relations modeled in the ontology, in order to connect different concepts. However, compared to the logic-based approach using the versatile Prolog representation, no simple specification/declaration of further processing knowledge like inference and derivation rules is possible. Likewise, in [66] mainly the instantiation of the knowledge elements is utilized in the mining process.

Typically, the knowledge graph mainly focuses on the structuring of the concepts and their relations, while specific modeling tasks, as well as data characteristics (e.g., distributions, correlations) are typically not captured. [13] presents a mixed-initiative approach, for semantic feature engineering using a knowledge graph. In a semi-automatic process, the knowledge graph is engineered and refined. Finally, the engineered features are provided for data mining. A similar approach is applied in [6]. Here, data from heterogeneous data sources is integrated into a knowledge graph, which then provides the basis for data mining by supporting feature selection, pattern mining, and interpretation in an integrated way. In particular, the constructed knowledge graph serves as a data integration and exploration mechanism, such that the modeled relations and additional information about the contained entities can be utilized by advanced graph mining methods, that work on such attributed graphs, e.g., [11].

Also, the obtained knowledge graph itself can be applied for providing additional context regarding the results of the data mining step, e.g., in order to provide explanations [10,83] as discussed below in more detail.

2.3   Explicative and Explanation-Aware Data Mining

The term explanation has been widely investigated in different disciplines. In this context, *explanation-aware* approaches have been a prominent research direction in artificial intelligence and data science, e.g., [10,44,73,91].

On Explanation. Knowing about kinds of explanations helps with structuring available knowledge and deciding which knowledge further is required for exhibiting certain explanation capabilities. In [74], Roth-Berghofer and Cassens outline the combination of goals and kinds of explanations, in the context of case-based reasoning. In [80], several useful kinds of explanations are discussed in the context of knowledge-based systems, referring to *concept explanations, purpose explanations, why explanations, action explanations,* and *how explanations,* cf., [10,80] for a detailed discussion.

In the data mining context, *concept, why* and *how* explanations are then particularly useful, since they provide insights into knowledge elements utilized in modeling, and also in the model itself by explicating model mechanisms and outcomes. Explanation goals, on the other hand, help to focus on user needs and expectations towards explanations. They aim at addressing to understand what and when the system has to be able to explain (something). Sørmo et al. [77] suggest a set of explanation goals addressing transparency, justification, relevance, conceptualisation, and learning.

Explicative Modeling. Recently, the concept of transparent and explainable models has also gained a strong focus and momentum in the data mining and machine learning community, e.g., [15,50,68], also see [32] for a survey on explaining black box models. Several methods focus on specific model types, e.g., tree-based models [84] or pattern-based approaches [26] for getting a better understanding of where a classifier does not work using local pattern mining techniques. Here, also methods integrating associative classification, i.e., based utilizing a set of (class) association rules [5,49,53,81] can be applied for obtaining interpretable models for explicative data mining. While the methods sketched above focus on specific modeling methods, there are several approaches for model agnostic explanation methods, e.g., [67,69]. In particular, general directions are given by methods considering counterfactual explanation, e.g., [55,90]. Furthermore, other general methods consider data perturbation and randomization techniques as well as interaction analysis methods, e.g., [33–35].

In general, for explicative data mining, the transparency of the respective patterns and models and their explanation-awareness is an important factor for supporting the user. In particular, if explanations for the complete models, or parts thereof can be provided, then the acceptance of the patterns and models, as well as their assessment and evaluation can often be significantly improved, e.g., [10].

Explanation-Aware Data Mining. The generation of explanations in the general data mining process is described in [10], the *mining and analysis continuum of explanation*. In particular, if explanations for the complete models, or parts thereof can be provided, then the acceptance can often be significantly improved, e.g., [10]. As put forward and described in the *Mining and Analysis Continuum of Explaining* [10] appropriate data representation and abstraction can facilitate explanation-awareness, also supporting and featuring different analysis and presentation levels. Then, data and models can be inspected at different levels of detail, from aggregated representations to the original ones in drill-down fashion combined with appropriate explanation capabilities. Typically, the user starts on an aggregated view that can be refined subsequently, for getting insights into the relations in the data and the constructed model, respectively. Here, different dimensions provide distinct view on the explanation space. Figure 3 provides an overview on the explanation framework and its dimensions. For a detailed discussion, we refer to [10].

In [4], for example, we consider symbolic representations, i.e., decision tree models and sequence representations of time series given by the symbolic aggregate approximation (SAX) [51,52] as a convenient data abstraction. In a general process model for explanation-aware data analytics, we investigate this abstraction together with a decision tree model in the context of feature selection and assessment, and present a case study in a petro-chemical production context.

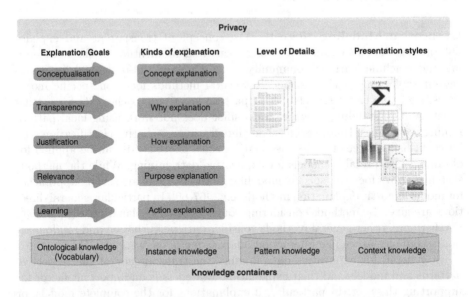

Fig. 3. Overview on the explanation dimensions of the mining and analysis continuum of explanation cf., [10].

Some recent approaches for introducing declarativity in explanation-aware approaches, include the knowledge-graph-based data mining approach outlined in [6] which is detailed in [11] regarding the applied pattern mining techniques. Furthermore, linked open data inspired approaches for interpreting pattern-based models, e.g., [62,63] and also explanations using linked open data for recommender systems [61] are first promising starting points in that context.

3   Conclusions

In this paper, we have provided an overview on declarative aspects in explicative data mining, targeting the overall goal of computational sensemaking. We have discussed the modeling of domain knowledge as well as extended knowledge structuring using knowledge graphs. Furthermore, we have summarized the paradigm of explicative data mining providing interpretable, transparent and explainable approaches.

We have introduced *explicative data mining* as a comprehensive paradigm. Similar to the philosophical process of *explication* cf., [19,54] which aims to make the implicit explicit, explicative data mining aims to model, describe and explain the underlying structure in the data. In that way, this paves the way to *computational sensemaking* which focuses on computational methods and models for "making sense" of complex data and information. Here, the goal is to understand structures and processes and to provide intelligent decision support through analysis and (semantic) interpretation. Therefore, explicative data mining coupled with declarative approaches is crucial since this can then both provide the necessary means for comprehensive analysis and giving meaning to models and results, respectively.

References

1. Agrawal, R., Srikant, R.: Fast algorithms for mining association rules. In: Proceedings of VLDB, pp. 487–499. Morgan Kaufmann (1994)
2. Atzmueller, M.: Subgroup discovery. WIREs Data Min. Knowl. Discov. 5(1), 35–49 (2015)
3. Atzmueller, M.: Onto explicative data mining: exploratory, interpretable and explainable analysis. In: Proceedings of Dutch-Belgian Database Day. TU Eindhoven, Netherlands (2017)
4. Atzmueller, M., Hayat, N., Schmidt, A., Klöpper, B.: Explanation-aware feature selection using symbolic time series abstraction: approaches and experiences in a petro-chemical production context. In: Proceedings of IEEE INDIN. IEEE Press, Boston (2017)
5. Atzmueller, M., Hayat, N., Trojahn, M., Kroll, D.: Explicative human activity recognition using adaptive association rule-based classification. In: Proceedings of IEEE International Conference on Future IoT Technologies. IEEE Press, Boston (2018, accepted)
6. Atzmueller, M., et al.: Big data analytics for proactive industrial decision support: approaches & first experiences in the context of the FEE project. atp edition 58(9) (2016)
7. Atzmueller, M., Lemmerich, F.: A methodological approach for the effective modeling of Bayesian networks. In: Dengel, A.R., Berns, K., Breuel, T.M., Bomarius, F., Roth-Berghofer, T.R. (eds.) KI 2008. LNCS (LNAI), vol. 5243, pp. 160–168. Springer, Heidelberg (2008). https://doi.org/10.1007/978-3-540-85845-4_20
8. Atzmueller, M., Puppe, F.: A methodological view on knowledge-intensive subgroup discovery. In: Staab, S., Svátek, V. (eds.) EKAW 2006. LNCS (LNAI), vol. 4248, pp. 318–325. Springer, Heidelberg (2006). https://doi.org/10.1007/11891451_28
9. Atzmueller, M., Puppe, F., Buscher, H.P.: Exploiting background knowledge for knowledge-intensive subgroup discovery. In: Proceedings of 19th International Joint Conference on Artificial Intelligence (IJCAI 2005), Edinburgh, Scotland, pp. 647–652 (2005)
10. Atzmueller, M., Roth-Berghofer, T.: The mining and analysis continuum of explaining uncovered. In: Proceedings of AI (2010)
11. Atzmueller, M., Schmidt, A., Kloepper, B., Arnu, D.: HYPGRAPHS: an approach for analysis and assessment of graph-based and sequential hypotheses. In: Appice, A., Ceci, M., Loglisci, C., Masciari, E., Raś, Z.W. (eds.) NFMCP 2016. LNCS (LNAI), vol. 10312, pp. 231–247. Springer, Cham (2017). https://doi.org/10.1007/978-3-319-61461-8_15

12. Atzmueller, M., Seipel, D.: Using declarative specifications of domain knowledge for descriptive data mining. In: Seipel, D., Hanus, M., Wolf, A. (eds.) INAP/WLP-2007. LNCS (LNAI), vol. 5437, pp. 149–164. Springer, Heidelberg (2009). https://doi.org/10.1007/978-3-642-00675-3_10

13. Atzmueller, M., Sternberg, E.: Mixed-initiative feature engineering using knowledge graphs. In: Proceedings of K-CAP. ACM (2017)

14. Baumeister, J., Atzmüller, M., Puppe, F.: Inductive learning for case-based diagnosis with multiple faults. In: Craw, S., Preece, A. (eds.) ECCBR 2002. LNCS (LNAI), vol. 2416, pp. 28–42. Springer, Heidelberg (2002). https://doi.org/10.1007/3-540-46119-1_4

15. Biran, O., Cotton, C.: Explanation and justification in machine learning: a survey. In: IJCAI 2017, Workshop on Explainable AI (2017)

16. Bizer, C., et al.: DBpedia - a crystallization point for the web of data. Web Semant. **7**(3), 154–165 (2009)

17. Blockeel, H.: Data mining: from procedural to declarative approaches. New Gener. Comput. **33**(2), 115–135 (2015)

18. Blockeel, H.: Declarative data analysis. Int. J. Data Sci. Anal., 1–7 (2017)

19. Carnap, R.: Logical Foundations of Probability (1962)

20. Chapman, P., et al.: CRISP-DM 1.0: Step-by-Step Data Mining Guide. CRISP-DM consortium: NCR Systems Engineering, DaimlerChrysler AG, SPSS Inc. and OHRA Verzekeringen en Bank Groep B.V (2000)

21. De Raedt, L., Kersting, K.: Statistical relational learning. In: Sammut, C., Webb, G.I. (eds.) Encyclopedia of Machine Learning, pp. 916–924. Springer, Boston (2011). https://doi.org/10.1007/978-0-387-30164-8_786

22. Dou, D., Wang, H., Liu, H.: Semantic data mining: a survey of ontology-based approaches. In: IEEE ICSC, pp. 244–251. IEEE (2015)

23. Duivesteijn, W., Knobbe, A., Feelders, A., van Leeuwen, M.: Subgroup discovery meets Bayesian networks-an exceptional model mining approach. In: Proceedings of International Conference on Data Mining (ICDM), pp. 158–167. IEEE, Washington, DC (2010)

24. Duivesteijn, W., Feelders, A., Knobbe, A.J.: Different slopes for different folks: mining for exceptional regression models with Cook's distance. In: Proceedings of ACM SIGKDD International Conference on Knowledge Discovery and Data Mining, pp. 868–876. ACM, New York (2012)

25. Duivesteijn, W., Feelders, A.J., Knobbe, A.: Exceptional model mining. Data Min. Knowl. Disc. **30**(1), 47–98 (2016)

26. Duivesteijn, W., Thaele, J.: Understanding where your classifier does (Not) work - the SCaPE model class for EMM. In: Proceedings of ICDM, pp. 809–814. IEEE (2014)

27. Fayyad, U.M., Piatetsky-Shapiro, G., Smyth, P.: From data mining to knowledge discovery: an overview. In: Advances in Knowledge Discovery and Data Mining, pp. 1–34. AAAI Press (1996)

28. Gamberger, D., Lavrac, N., Wettschereck, D.: Subgroup visualization: a method and application in population screening. In: Proceedings of IDAMAP (2002)

29. Gaskin, C.J., Happell, B.: On exploratory factor analysis: a review of recent evidence, an assessment of current practice, and recommendations for future use. Int. J. Nurs. Stud. **51**(3), 511–521 (2014)

30. Goethals, B., Moens, S., Vreeken, J.: MIME: a framework for interactive visual pattern mining. In: Proceedings of ACM SIGKDD, pp. 757–760. ACM (2011)

31. Goodman, B., Flaxman, S.: European union regulations on algorithmic decision-making and a "right to explanation". arXiv preprint arXiv:1606.08813 (2016)

32. Guidotti, R., Monreale, A., Turini, F., Pedreschi, D., Giannotti, F.: A survey of methods for explaining black box models. arXiv preprint arXiv:1802.01933 (2018)
33. Henelius, A., Puolamäki, K., Boström, H., Asker, L., Papapetrou, P.: A peek into the black box: exploring classifiers by randomization. Data Min. Knowl. Discov. **28**(5–6), 1503–1529 (2014)
34. Henelius, A., et al.: GoldenEye++: a closer look into the black box. In: Gammerman, A., Vovk, V., Papadopoulos, H. (eds.) SLDS 2015. LNCS (LNAI), vol. 9047, pp. 96–105. Springer, Cham (2015). https://doi.org/10.1007/978-3-319-17091-6_5
35. Henelius, A., Puolamäki, K., Ukkonen, A.: Interpreting classifiers through attribute interactions in datasets. In: Proceedings of 2017 ICML Workshop on Human Interpretability in Machine Learning (WHI 2017) (2017)
36. Hoffart, J., Suchanek, F.M., Berberich, K., Weikum, G.: YAGO2: a spatially and temporally enhanced knowledge base from Wikipedia. Artif. Intell. **194**, 28–61 (2013)
37. Jaroszewicz, S., Simovici, D.A.: Interestingness of frequent itemsets using Bayesian networks as background knowledge. In: Proceedings of SIGKDD, pp. 178–186. ACM (2004)
38. Kaytoue, M., Plantevit, M., Zimmermann, A., Bendimerad, A., Robardet, C.: Exceptional contextual subgraph mining. Mach. Learn. **106**(8), 1171–1211 (2017)
39. Keim, D., Ward, M.: Visualization. In: Berthold, M., Hand, D.J. (eds.) Intelligent Data Analysis. Springer, Heidelberg (2003). https://doi.org/10.1007/978-3-540-48625-1_11
40. Klösgen, W.: Explora: a multipattern and multistrategy discovery assistant. In: Advances in Knowledge Discovery and Data Mining, pp. 249–271. AAAI Press (1996)
41. Klösgen, W.: Subgroup discovery. In: Handbook of Data Mining and Knowledge Discovery. Oxford University Press, New York (2002). Chap. 16.3
42. Klösgen, W., Lauer, S.R.W.: Visualization of data mining results. In: Handbook of Data Mining and Knowledge Discovery. Oxford University Press, New York (2002). Chap. 20.1
43. Knobbe, A.J., Cremilleux, B., Fürnkranz, J., Scholz, M.: From local patterns to global models: the LeGo approach to data mining. In: From Local Patterns to Global Models: Proceedings of the ECML/PKDD-08 Workshop (LeGo 2008), pp. 1–16 (2008)
44. Kolodner, J.L.: Making the implicit explicit: clarifying the principles of case-based reasoning. In: Case-based Reasoning: Experiences, Lessons and Future Directions, pp. 349–370 (1996)
45. Lavrač, N.: Subgroup discovery techniques and applications. In: Ho, T.B., Cheung, D., Liu, H. (eds.) PAKDD 2005. LNCS (LNAI), vol. 3518, pp. 2–14. Springer, Heidelberg (2005). https://doi.org/10.1007/11430919_2
46. Lavrac, N., Kavsek, B., Flach, P., Todorovski, L.: Subgroup discovery with CN2-SD. J. Mach. Learn. Res. **5**, 153–188 (2004)
47. Leman, D., Feelders, A., Knobbe, A.: Exceptional model mining. In: Daelemans, W., Goethals, B., Morik, K. (eds.) ECML PKDD 2008. LNCS (LNAI), vol. 5212, pp. 1–16. Springer, Heidelberg (2008). https://doi.org/10.1007/978-3-540-87481-2_1
48. Lemmerich, F., Becker, M., Atzmueller, M.: Generic pattern trees for exhaustive exceptional model mining. In: Flach, P.A., De Bie, T., Cristianini, N. (eds.) ECML PKDD 2012. LNCS (LNAI), vol. 7524, pp. 277–292. Springer, Heidelberg (2012). https://doi.org/10.1007/978-3-642-33486-3_18

49. Li, W., Han, J., Pei, J.: CMAR: accurate and efficient classification based on multiple class-association rules. In: Cercone, N., Lin, T.Y., Wu, X. (eds.) Proceedings of International Conference on Data Mining (ICDM), pp. 369–376. IEEE Computer Society (2001)

50. Li, X., Huan, J.: Constructivism learning: a learning paradigm for transparent predictive analytics. In: Proceedings of SIGKDD, pp. 285–294. ACM (2017)

51. Lin, J., Keogh, E., Lonardi, S., Chiu, B.: A symbolic representation of time series, with implications for streaming algorithms. In: Proceedings of 8th ACM SIGMOD Workshop on Research Issues in Data Mining and Knowledge Discovery, pp. 2–11. ACM (2003)

52. Lin, J., Keogh, E., Wei, L., Lonardi, S.: Experiencing SAX: a novel symbolic representation of time series. Data Min. Knowl. Discov. **15**(2), 107–144 (2007)

53. Liu, B., Hsu, W., Ma, Y.: Integrating classification and association rule mining. In: Proceedings of SIGKDD, pp. 80–86. AAAI Press, August 1998

54. Maher, P.: Explication defended. Studia Logica **86**(2), 331–341 (2007)

55. Mandel, D.R.: Counterfactual and causal explanation: from early theoretical views to new frontiers. In: The Psychology of Counterfactual Thinking, pp. 23–39. Routledge (2007)

56. Mitzlaff, F., Atzmueller, M., Hotho, A., Stumme, G.: The social distributional hypothesis. J. Soc. Netw. Anal. Min. **4**(216), 1–14 (2014)

57. Mitzlaff, F., Atzmueller, M., Stumme, G., Hotho, A.: Semantics of user interaction in social media. In: Ghoshal, G., Poncela-Casasnovas, J., Tolksdorf, R. (eds.) Complex Networks IV. SCI, vol. 476. Springer, Heidelberg (2013). https://doi.org/10.1007/978-3-642-36844-8_2

58. Morik, K.: Detecting interesting instances. In: Hand, D.J., Adams, N.M., Bolton, R.J. (eds.) Pattern Detection and Discovery. LNCS (LNAI), vol. 2447, pp. 13–23. Springer, Heidelberg (2002). https://doi.org/10.1007/3-540-45728-3_2

59. Morik, K., Boulicaut, J.-F., Siebes, A. (eds.): Local Pattern Detection. LNCS (LNAI), vol. 3539. Springer, Heidelberg (2005). https://doi.org/10.1007/b137601

60. Morshed, A., Dutta, R., Aryal, J.: Recommending environmental knowledge as linked open data cloud using semantic machine learning. In: Proceedings of IEEE ICDEW, pp. 27–28. IEEE (2013)

61. Musto, C., Narducci, F., Lops, P., de Gemmis, M., Semeraro, G.: Linked open data-based explanations for transparent recommender systems. Int. J. Hum.-Comput. Stud. (2018)

62. Paulheim, H.: Explain-a-LOD: using linked open data for interpreting statistics. In: Proceedings of ACM IUI, pp. 313–314. ACM (2012)

63. Paulheim, H.: Generating possible interpretations for statistics from linked open data. In: Simperl, E., Cimiano, P., Polleres, A., Corcho, O., Presutti, V. (eds.) ESWC 2012. LNCS, vol. 7295, pp. 560–574. Springer, Heidelberg (2012). https://doi.org/10.1007/978-3-642-30284-8_44

64. Paulheim, H., Fümkranz, J.: Unsupervised generation of data mining features from linked open data. In: Proceedings of WIMS, p. 31. ACM (2012)

65. Pujara, J., Miao, H., Getoor, L., Cohen, W.: Large-scale knowledge graph identification using PSL. In: AAAI Fall Symposium on Semantics for Big Data (2013)

66. Rauch, J., Šimůnek, M.: Learning association rules from data through domain knowledge and automation. In: Bikakis, A., Fodor, P., Roman, D. (eds.) RuleML 2014. LNCS, vol. 8620, pp. 266–280. Springer, Cham (2014). https://doi.org/10.1007/978-3-319-09870-8_20

67. Ribeiro, M.T., Singh, S., Guestrin, C.: Model-agnostic interpretability of machine learning. In: Proceedings of 2016 ICML Workshop on Human Interpretability in Machine Learning (2016)
68. Ribeiro, M.T., Singh, S., Guestrin, C.: Why should i trust you? Explaining the predictions of any classifier. In: Proceedings of ACM SIGKDD, pp. 1135–1144. ACM (2016)
69. Ribeiro, M.T., Singh, S., Guestrin, C.: Anchors: high-precision model-agnostic explanations. AAAI (2018)
70. Richardson, M., Domingos, P.: Learning with knowledge from multiple experts. In: Proceedings of ICML, pp. 624–631. AAAI Press (2003)
71. Richardson, M., Domingos, P.: Markov logic networks. Mach. Learn. **62**(1–2), 107–136 (2006)
72. Ristoski, P., Paulheim, H.: Semantic web in data mining and knowledge discovery: a comprehensive survey. Web Semant. **36**, 1–22 (2016)
73. Roth-Berghofer, T., Schulz, S., Leake, D., Bahls, D.: Explanation-aware computing. AI Mag. **28**(4) (2007)
74. Roth-Berghofer, T.R., Cassens, J.: Mapping goals and kinds of explanations to the knowledge containers of case-based reasoning systems. In: Muñoz-Ávila, H., Ricci, F. (eds.) ICCBR 2005. LNCS (LNAI), vol. 3620, pp. 451–464. Springer, Heidelberg (2005). https://doi.org/10.1007/11536406_35
75. Seipel, D., Nogatz, F., Abreu, S.: Domain-specific languages in prolog for declarative expert knowledge in rules and ontologies. Comput. Lang. Syst. Struct. **51**, 102–117 (2018)
76. Shneiderman, B.: The eyes have it: a task by data type taxonomy for information visualizations. In: Proceedings of IEEE Symposium on Visual Languages, Boulder, Colorado, pp. 336–343 (1996)
77. Sørmo, F., Cassens, J., Aamodt, A.: Explanation in case-based reasoning - perspectives and goals. Artif. Intell. Rev. **24**(2), 109–143 (2005)
78. Spenke, M.: Visualization and interactive analysis of blood parameters with Info-Zoom. Artif. Intell. Med. **22**(2), 159–172 (2001)
79. Spenke, M., Beilken, C.: Visual, interactive data mining with InfoZoom - the financial dataset. In: Workshop Notes on Discovery Challenge at the 3rd European Conference on Principles and Practice of Knowledge Discovery in Databases, pp. 15–18 (1999)
80. Spieker, P.: Natürlichsprachliche Erklärungen in technischen Expertensystemen. Dissertation, University of Kaiserslautern (1991)
81. Thabtah, F.: A review of associative classification mining. Knowl. Eng. Rev. **22**(1), 37–65 (2007)
82. Theus, M.: Interactive data visualization using Mondrian. J. Stat. Softw. **7**(11), 1–9 (2003)
83. Tiddi, I., d'Aquin, M., Motta, E.: An ontology design pattern to define explanations. In: Proceedings of K-Cap. ACM, New York (2015)
84. Tolomei, G., Silvestri, F., Haines, A., Lalmas, M.: Interpretable predictions of tree-based ensembles via actionable feature tweaking. In: Proceedings of the 23rd ACM SIGKDD International Conference on Knowledge Discovery and Data Mining, pp. 465–474. ACM (2017)
85. Van Deursen, A., Klint, P., Visser, J.: Domain-specific languages: an annotated bibliography. ACM Sigplan Not. **35**(6), 26–36 (2000)

86. Leeuwen, M.: Interactive data exploration using pattern mining. In: Holzinger, A., Jurisica, I. (eds.) Interactive Knowledge Discovery and Data Mining in Biomedical Informatics. LNCS, vol. 8401, pp. 169–182. Springer, Heidelberg (2014). https://doi.org/10.1007/978-3-662-43968-5_9

87. Vavpetič, A., Lavrač, N.: Semantic subgroup discovery systems and workflows in the SDM-Toolkit. Comput. J. **56**(3), 304–320 (2013)

88. Vavpetic, A., Podpecan, V., Lavrac, N.: Semantic subgroup explanations. J. Intell. Inf. Syst. **42**(2), 233–254 (2014)

89. Velicer, W.F., Eaton, C.A., Fava, J.L.: Construct explication through factor or component analysis: a review and evaluation of alternative procedures for determining the number of factors or components. In: Goffin, R.D., Helmes, E. (eds.) Problems and Solutions in Human Assessment: Honoring Douglas N. Jackson at Seventy, pp. 41–71. Springer, Boston (2000). https://doi.org/10.1007/978-1-4615-4397-8_3

90. Wachter, S., Mittelstadt, B., Russell, C.: Counterfactual Explanations without Opening the Black Box: Automated Decisions and the GDPR (2017)

91. Wick, M.R., Thompson, W.B.: Reconstructive expert system explanation. Artif. Intell. **54**(1–2), 33–70 (1992)

92. Wilcke, X., Bloem, P., de Boer, V.: The knowledge graph as the default data model for learning on heterogeneous knowledge. Data Sci. (Preprint), 1–19 (2017)

93. Wirth, R., Hipp, J.: CRISP-DM: towards a standard process model for data mining. In: Proceedings of 4th International Conference on the Practical Application of Knowledge Discovery and Data Mining, pp. 29–39. Morgan Kaufmann (2000)

94. Wrobel, S.: An algorithm for multi-relational discovery of subgroups. In: Komorowski, J., Zytkow, J. (eds.) PKDD 1997. LNCS, vol. 1263, pp. 78–87. Springer, Heidelberg (1997). https://doi.org/10.1007/3-540-63223-9_108

95. Zelezny, F., Lavrac, N., Dzeroski, S.: Using constraints in relational subgroup discovery. In: Proceedings of International Conference on Methodology and Statistics, pp. 78–81. University of Ljubljana (2003)

An Approach for Representing Answer Sets in Natural Language

Min Fang and Hans Tompits[✉]

Institute of Logic and Computation, Knowledge-Based Systems Group E192-03,
Technische Universität Wien, Favoritenstraße 9-11, 1040 Vienna, Austria
{fang,tompits}@kr.tuwien.ac.at

Abstract. In recent years, different methods for supporting the development of answer-set programming (ASP) code have been introduced. During such a development process, often it would be desirable to have a natural-language representation of answer sets, e.g., when dealing with domain experts unfamiliar with ASP. In this paper, we address this point and provide an approach for such a representation, defined in terms of a controlled natural language (CNL), which in turn relies on the annotation language LANA for the specification of meta-information for answer-set programs. Our approach has been implemented as an Eclipse plug-in for SeaLion, a dedicated IDE for ASP.

1 Introduction

In recent years, the question of providing methods and tools for supporting the development of answer-set programs has received increased attention in the literature [2,8]. A feature often desirable in developing answer-set programs is to have a natural-language representation of the output of such programs, i.e., of answer sets which represent the solutions of encoded problems. Such representations are particularly useful, e.g., when dealing with domain experts who are unfamiliar with answer-set programming (ASP).

In this paper, we address this point and provide an approach for such a natural-language representation of answer sets. However, for realising a method like this, it should be clear that information other than the pure ASP code is required. Somehow, a mechanism which allows to describe what certain ASP predicates "mean" is required. Such a feature is provided by the annotation language LANA ("Language for ANnotating Answer-set programs") [3] that augments answer-set programs with additional meta-information. LANA offers language constructs for declaring various concepts, like predicates, type information, and input and output signatures. These additional annotations can then be interpreted by independent tools to support the development process, e.g., to run specialised test cases, to automatically generate a program documentation, and so on. In fact, LANA, and some of the tools using it, have already been integrated in SeaLion [2], a dedicated integrated development environment (IDE) for ASP implemented as an Eclipse plug-in.

© Springer Nature Switzerland AG 2018
D. Seipel et al. (Eds.): DECLARE 2017, LNAI 10997, pp. 115–131, 2018.
https://doi.org/10.1007/978-3-030-00801-7_8

Our goal was to create a tool which uses the information provided by LANA and which generates a more human-readable interpretation of answer sets. For this purpose, we used the user-specified atom descriptions (given by means of LANA @atom annotations) and built natural-language-like interpretations for the answer sets according to the problem instance. In order to parse the atom descriptions properly and deduce syntactic information from them, our approach uses descriptions which are formulated in a *controlled natural language* (CNL). Our tool, realised as an Eclipse plug-in alongside SeaLion, allows a knowledge engineer to make further ad-hoc changes to a generated interpretation and eventually export it, e.g., in PDF format. This document can then be easily forwarded to a domain expert for further consideration.

Following Kuhn [15], a CNL is a constructed language resembling a natural language but being more restrictive concerning lexicon, syntax, and/or semantics while preserving most of its natural properties. In effect, it is essentially a formal language and can therefore usually be specified by known mechanisms for specifying formal languages, e.g., in terms of a grammar. The degree of resemblance of a CNL to a natural language varies depending on the purpose of the CNL and the (technical) background of the designated users of this language. A CNL is often used as a bridge between a highly ambiguous natural language and a less human-readable formal language, like, e.g., predicate logic or a programming language. This type of CNL is referred to as *computer-oriented CNL* and is usually employed in applications where some sort of semi-automatic or automatic processing of user input is necessary in order to translate it into a more formal representation [21]. Recent instances of general-purpose (i.e., not restricted to the vocabulary of a specific domain), English-based CNLs are, e.g., Attempto Controlled English (ACE) [9,10] and PENG/PENG Light [19,20,24].

Our approach is not the first to employ CNL techniques in the context of answer-set programming. To wit, Erdem and Yeniterzi [6] developed BioQuery-CNL, a domain-specific CNL which is used to express biomedical queries over predefined ontologies. These queries are then in turn translated into answer-set programs, using the BioQuery-ASP system [4,5]. Likewise, Schwitter [22,23] as well as Guy and Schwitter [12] discuss methods to solve search problems by representing them into a CNL and processing these representations by translations to ASP.

The paper is organised as follows: We first provide, in Sect. 2, some background on the ASP annotation language LANA. Then, in Sect. 3, we describe our proposed CNL for obtaining a natural-language representation of answer sets and Sect. 4 details the actual translation. Section 5 briefly discusses the implemented tool and the paper closes with some general remarks in Sect. 6.

2   Background

We assume the reader familiar with the basics of answer-set programming (ASP) [1]. Briefly, an *answer-set program* consists of a set of rules of form

$$a \; :- \; b1, \ldots, bn, not \; c1, \ldots, not \; cm,$$

where a, b1,..., bn, c1,..., cm are atoms over a first-order vocabulary and not denotes *default negation*. The semantics of such a program, P, is given in terms of *answer sets*, which are defined as those models of P which satisfy a certain fixed-point condition [11]. Prominent solvers for computing answer sets are, e.g., clasp [18] and DLV [16].

```
%**
@block WorkTable {
assign projects and jobs to employees

@atom employee(E)
E is an employee.

@atom skill(S)
S is a skill.

@atom project(P)
P is a project.

@atom hasSkill(E,S)
Employee E has skill S.

@atom requiresSkill(P,S).
Project P requires skill S.

@atom works(E,P,S)
Employee E works on project P with skill(s) S.

@atom projectLeader(P,E)
Employee E is project leader for project P.

@input employee/1, skill/1, project/1, hasSkill/2,
        requiresSkill/2

@output works/3, projectLeader/2

*%

% ASP rules feeding the grounder/solver

...

%**

} % end of block WorkTable

*%
```

Fig. 1. An illustration of LANA annotations in a program file.

The annotation language LANA was introduced by De Vos et al. [3] with the purpose of defining a standardised apparatus for specifying meta-information for answer-set programs. The formalism of LANA is reminiscent of Java annotations, with the @ symbol preceding each keyword. Just as some Java annotations (e.g., annotations for Javadoc), annotations in LANA have the form of an ASP comment, thus not altering the semantics of the program that it is documenting. If an ASP solver—in particular its grounding component—supports block-comments, these can be used instead of the single-line comment marker "%".[1]

LANA offers an array of different annotations, of which we only mention a few. E.g., @block can be used to group certain rules together, i.e., deriving a more finely-grained structure within a code file. Each block can then declare predicates using the @atom annotation, and define its input and output signature with @input and @output, respectively. Other annotations include @assert, @precon, @postcon, @always, and @never, which define logical conditions for the answer sets and can be used for testing purposes.

For our use case, we mainly focus on two of the annotations: @atom and @output. Figure 1 depicts some annotations for illustrating the usage of these two annotations. As can be seen in this example, @atom annotations are made up of a predicate name, a comma-separated list of its arguments given as variables (i.e., an identifier starting with an uppercase character) wrapped in round brackets, and an (optional) sentence describing the semantics of the corresponding predicate. LANA itself does not define any restrictions for this short description as it is only considered a comment, i.e., it is used as an informal documentation of the associated predicate. For our purposes, however, it is necessary to impose a certain well-defined structure on them.

The @input and @output annotations list the predicates that correspond to the input and output of a problem instance, respectively. Input predicates are usually required to encode the problem instance, while output predicates are relevant for the solution of a particular problem instance. The predicates are given with their name and their respective arity since it is possible to have predicates with the same name, in which case they are uniquely identifiable by the combination of their name and arity.

While in LANA all these annotations are not mandatory, we presume for our purposes that they are specified by the user, otherwise it would be impossible to generate a textual, human-readable interpretation of answer sets that would reflect their intended meaning.

---

[1] E.g., if one uses gringo as grounding component. An additional "*" is then added to the block-comment marker "%*" in order to distinguish LANA annotations from normal block comments. Hence, LANA annotations are wrapped in "%**" and "*%" blocks. We assume this syntax for the examples below.

3   A Controlled Natural Language for LANA Atom Descriptions

We first define the controlled natural language that determines the form of atom descriptions associated with an @atom annotation. The CNL that we are envisaging for the LANA atom descriptions cannot be domain-specific since we are not restricting the domain of answer-set programs that can use the LANA annotations (e.g., it is possible to encode a combinatoric puzzle as much as a scheduling problem). Hence, it is impossible to predefine a set lexicon for our CNL as far as content words are concerned. We can, however, presume a fixed set of function words (e.g., prepositions) which we incorporate directly into the grammar.

3.1   Preliminary Considerations

Having an open, utterly unconstrained lexicon is unusual for both domain-specific and general-purpose CNLs. The former would normally have a relatively small lexicon tailored to its field of application. For the latter, a dynamic lexicon is more typical, i.e., it would have a compact predefined lexicon with the most frequently used words of the associated natural language, with the additional possibility of the user specifying new entries for the lexicon as they go along. Even though a dynamic lexicon could be used for our purposes, it would complicate the user experience unnecessarily since the user may have to specify further information about the new lexemes (e.g., identifying their part-of-speech).

The real restriction of our CNL therefore lies in the syntactic structure of its sentences. Instead of a hundred odd rules (e.g., ACE Codeco has 164 grammar rules [15]), we constrain our grammar to a very small number of rules. This is possible because our CNL is, after all, not a general-purpose CNL. We know its application context (stated below an @atom annotation describing the meaning of this atom) and its purpose (generating human-readable interpretations of answer sets) despite being uninformed about the semantics of its individual words and the domain of the program. By imposing a rigid structure on the sentences we gain enough structural information about these sentences and their constituents that we can use for the generation of new sentences, i.e., textual interpretations, according to the instantiated atoms of each answer set.

This highly syntactic approach is very much different from some of the more well-known CNLs, which usually map the sentences to some kind of formal logic representation (e.g., to discourse representation structures [10,24]). We are able to bypass this step since we are not so much interested in the exact semantics of the sentences, but rather in the structural relations of its constituents. Indeed, we use these constituents as "boilerplates" for the generation of our own sentences, which make up the textual interpretation of answer sets.

Concerning the grammatical structure that we are defining for LANA atom descriptions, the first general restriction that we make is that each atom description is made up of one sentence only. This way we can avoid difficult problems such as anaphora resolution.[2] The structure of these sentences must be basic and unambiguously parseable but at the same time flexible enough to allow the description of many possible predicates. We essentially allow three sentence types in our CNL, where the defining characteristics are linked to the main verb of the sentence and its argument structure, i.e., the number and the kind of arguments it combines with. In traditional grammar, this verb property is referred to as *valency* or *valence* of verbs (cf., e.g., Van Valin [26]).

3.2 Syntactical Structure of the CNL

We now detail the exact constituents each sentence type of our CNL allows and what each constituent is made up of. Instead of giving a general formal definition of the syntax of our CNL, which we omit here due to space reasons, we focus on providing illustrative explanations what valid phrases and sentences our language admits (for a full specification of the syntax of our CNL, cf. Fang [7]).

For our purposes it is sufficient that we include a fixed set of (more or less) function words that our parser recognises and which help to structure the input. In general, function words belong to the *closed-word class*, i.e., this set of words typically does not grow larger with time (like, e.g., nouns and adjectives do). Hence, we could theoretically exhaustively include the set of function words into our grammar, but to stay in line with our goal to be as restrictive as possible with the defined grammar without giving up too much expressive power, we focus on a limited set of function words that seem to cover our applications well enough. This final set is made up of the following words: *is, are, be, has, have, do, does, must, can, cannot, an, a, A, An, the, The, There, maximally, minimally, at least, at most, or more, or less, or fewer, to, with, of, for, as, by, at, in, on,* and *not.* The prepositions are especially important since they help the parser to unambiguously place prepositional phrases in the sentence (and thus identify where a noun phrase ends). In the examples below, instances of these words are underlined.

Central Language Constituents. We first list central language elements of our CNL.

Variables. We have predefined variables of a specific form: They must start with an upper-case character, followed by other upper-case characters or numbers 0 to 9. Note that this definition is more restricted than the usual ASP definition for variables. There is, however, one exception to this simple rule: The symbol "A" cannot be used to denote a variable since it is a reserved function word (referring to an indefinite article at the beginning of a sentence). In the subsequent examples variables are printed in bold.

Noun Phrases (NPs). First of all, we distinguish between noun phrases (NPs) which contain variables identical to those listed in the @atom signature and noun

---

[2] PENG [24] avoids this problem by disallowing personal pronouns, which are often contextually ambiguous, and using explicit variable references instead.

phrases which do not. *Variable-free* NPs (NP^~~var~~) are made up of an arbitrary number of words that are not in the set of function words (only the first one may be a definite or indefinite article). Since the words may be chosen at random by the user, they do not necessarily have to be real noun phrases, they can in fact be adjectival phrases (AdjPs) too (however, we will keep the label NP because the labelling has little importance). Because there is almost no restriction within a variable-free NP and also because they can be difficult to parse depending on the surrounding constituents, their availability in our grammar is very much restricted (details are given further down). Typical NP^~~var~~ instances are the following:

(1)   a. <u>a</u> ship;
      b. battle ships;
      c. <u>the</u> best project leader.

Variable-containing NPs (NP^{var}) are in their most minimalistic form made up of the variable only. Optionally, they may then contain a definite or an indefinite article. In their most elaborate form they may additionally contain arbitrary words between the article and the variable (basically describing the semantics of the variable) and a post-variable modifier. Possible modifiers are *maximally*, *minimally*, *at least*, *at most*, *or more*, *or less*, and *or fewer*. Hence, examples for variable-containing NPs typically look like the following:

(2)   a. **E**;
      b. <u>an</u> employee **E**;
      c. <u>the</u> project leader **L**;
      d. <u>a</u> proficiency level **V** <u>at least</u>.

However, sometimes the position of the variable and the describing noun may be reversed (we use NP^rev to refer to this special case and NP^~~rev~~ to the complementary case). This is usually the case if the variable is of type integer. In these cases, the variable will be instantiated with integer constants in corresponding answer sets. A modifier is also possible in this type of NP after the last word. Note that there must be at least one word following the variable in order to determine the concept described or quantified by the variable. Examples of this type of NP are the following:

(3)   a. **N** employees;
      b. **N** employees <u>or more</u>.

As an additional feature, we enable the user to specify the plural suffix of one-word nouns so that this information can also be automatically used in the interpretation generation process. For instance, consider the following example:

(4)   a. **N** employee(s);
      b. project(s) **P**.

Hence, if desired, one can add the plural suffix "-s" to a noun in brackets. This addition will especially make sense in NP^{rev} cases. It will then be used in generated sentences where it is clear that a plural form of the marked noun will be required. If this additional information is not given by the user, the generation process will simply stick to the form given.

Prepositional Phrases (PPs). Prepositional phrases (PPs) are a combination of a preposition and an NP, where the NP can be variable-free or variable-containing. Hence, a PP is either PP^{var} or $\text{PP}^{\overline{\text{var}}}$ depending on the NP that they contain. Prepositions are comprised of the words *to, with, of, for, as, by, at, in,* and *on.* Typical examples are the following:

(5) a. by **E**;
  b. on project **P**;
  c. for **N** employees at most;
  d. as the project leader.

Sentence Types. We next describe our three categories of sentences. Note that we use a classification of verbs into *intransitive* (valency 1), *transitive* (valency 2), and *ditransitive* (valency 3) verbs.[3] Intransitive verbs come with a subject only, transitive verbs must have a subject and a direct object (DO), while ditransitive verbs combine a subject, a DO, and an indirect object (IO), which is normally realised as an oblique object (i.e., it is inside a preposition phrase) in English.

Sentences with an Intransitive Verb. The main characteristic of sentences with an intransitive verb is the fact that the verb only requires one argument, which is in subject position. Hence, we allow for this type of sentences only one NP, which furthermore must contain a variable. Consequently, a corresponding atom would have arity one or more. If it does indeed have a higher arity, the remaining variables must be encoded into PPs. Schematically, such sentences have the following form:[4]

(6) NP^{var} Modal? V^1 PP*.

As indicated here, we also allow the addition of a modal verb before the main verb, which is either *can* or *must.* Moreover, the negated version of an acceptable sentence is also acceptable (the accepted negation strategy depends on the verb configuration in the sentence). Some examples are the following:

(7) a. [Employee **E**] works.
  b. [Employee **E**] does not work.

---

[3] Verbs with valency 0 do not have their own term since there is only a small number of them (predominantly weather verbs). We disregard them for our considerations.

[4] In what follows, we use superscripts to denote the valency of the associated verb and the symbols "?" and "*" refer to BNF syntax customs (i.e., standing for options and possible repetitions, respectively).

    c. [**N** employee(s)] <u>cannot</u> work.

    d. [Employee **E1**] works [<u>with</u> employee **E2**] [<u>on</u> project **P**].

    e. [**N** employees] <u>must</u> work [<u>on</u> project **P**].

    f. [Employee **E**] <u>must not</u> work [<u>on</u> project **P**].

Note again that the first constituent must be an NP containing a variable. A variable-free NP in this position will lead to a parsing error.

Sentences with a Transitive Verb. Since a transitive verb requires two arguments, one in subject position and one in object position, this sentence type requires two NPs, both of which have to contain a variable corresponding to the associated atom signature. For the remaining variables declared by the atom, PPs should be used, which would wrap variable-containing NPs. Schematically, this kind of sentences have the following form:

(8)   NP^{var} Modal? V$^{2/3}$ NP^{var} PP*.

Again, an additional modal verb just before the main verb is possible as well as the negated version of allowed sentences, like in the following examples:

(9)   a. [Employee **E**] heads [project **P**].

      b. [Employee **E**] <u>does not</u> head [project **P**].

      c. [Employee **E**] heads [project **P**] [<u>with</u> skill **S**].

      d. [A ship **T**] <u>must</u> occupy [position **X**] [<u>on</u> day **D**].

      e. [A ship **T**] <u>must not</u> occupy [position **X**].

Sentences with a Copula. In linguistics, the term *copula* is used to refer to a certain kind of linking verb that connects the subject of a sentence to the subject complement, the so-called *predicative*. Informally, a copula can be understood to be similar to the mathematical "equals" sign, equating its left part to its right part. The main copula in English is the verb *to be*. Sentences with a copula allow the user more freedom with certain constituents. Contrary to the other types, it is possible to have both a variable-containing and a variable-free NP. In the first case, the copula can be followed by either an NP^{var} or an NP^{var} and optional PPs. Schematically, such sentences look as follows:[5]

(10)   NP^{var} Modal? V^{cop} NP PP*.

In the latter case, however, we have to make sure that the sentence still contains at least one variable. Since the first NP does not contain one, there are only two other options: Either there is an NP^{var} occurring immediately after the copula, in which case there are no further PPs necessary, or there is only an NP^{var} or no NP at all after the copula, in which case there must be at least one PP at the end of the sentence that contains an NP^{var}. This is schematically summarised as follows:

---

[5] By NP we denote the union of NP^{var} and NP^{var}. Similarly, PP denotes the union of PP^{var} and PP^{var}.

(11)   a. NP^~~var~~ Modal? V^cop NP^var PP*;

   b. NP^~~var~~ Modal? V^cop NP^~~var~~? PP^var PP*.

Let us illustrate schemata (10) and (11) with examples (12) and (13) below, respectively:

(12)   a. [Employee **E**] is [a project leader].

   b. [Employee **E**] is [project leader] [for project **P**].

   c. [Employee **E**] must not be [project leader] [for project **P**].

(13)   a. [An employee] is [project leader **L**] [for project **P**].

   b. [There] are [**N** project leader(s)] [on project **P**].

   c. [The Planes] are [in airport **X**] [with identifier **Y**].

   d. [Planes] cannot be [in airport **X**].

   e. [There] must be [planes] [in airport **X**] [at day **T**].

   f. [There] must not be [project leaders] [for project **P**].

Observe that by allowing variable-free NPs in the subject position, existential constructions with *there* become available.

4   Interpreting Answer Sets

Based on the @atom annotations provided by the LANA language in an ASP code and the natural-language descriptions of the atoms conforming to the CNL of the previous section, we can generate adapted sentences that make answer sets more readable.

```
works(boris,p2,planning) works(lisa,p2,planning)
works(boris,p2,marketing) works(boris,p2,design)
works(lisa,p2,design) works(lisa,p2,modelling)
works(sarah,p2,modelling) works(boris,p1,marketing)
works(boris,p1,design) works(sarah,p1,modelling)
works(peter,p1,planning) works(hans,p1,modelling)
works(sarah,p1,coding) works(peter,p1,coding)
projectLeader(p1,sarah) projectLeader(p2,lisa)
```

Fig. 2. A possible answer set of the program mentioned in Fig. 1

Basic Generation. Let us revisit the example given in Fig. 1. Suppose that this code forms part of a program that takes facts as input which encode information about employees and projects and is supposed to output answer sets that can be interpreted as possible project assignments. Following the ASP "guess and check" methodology, there would be a guessing part in the program generating candidate solutions and some rules and constraints which filter out those candidates which are not proper solutions of the problem instance.

A domain expert is usually only interested in those predicates which encode the problem solution. Hence, using the @output annotation of LANA, one can mark the relevant atoms that should be considered for the interpretation-generation process. If all predicates are relevant, this annotation may be left out since the default assumption is then that every predicate should be used.

Consider again the example from Fig. 1 and the answer set of the program mentioned therein depicted in Fig. 2, which has been stripped off of irrelevant atoms, i.e., atoms that were not marked as @output in the program code.

Using the information provided by the @atom annotation and the atom description directly below the annotation, we can generate interpretation sentences corresponding to this answer set by simply replacing each variable place holder in the sentence with the appropriate constant in the answer set atom. This basic method will generate a new sentence for each instantiated atom in the given answer set. The result of this naive procedure is as follows:

$$\text{Employee boris works on project p1 with skill design.} \tag{14}$$

$$\text{Employee boris works on project p1 with skill marketing.} \tag{15}$$

$$\text{Employee boris works on project p2 with skill design.} \tag{16}$$

$$\text{Employee boris works on project p2 with skill marketing.} \tag{17}$$

$$\text{Employee boris works on project p2 with skill planning.} \tag{18}$$

$$\text{Employee hans works on project p1 with skill modelling.} \tag{19}$$

$$\text{Employee lisa works on project p2 with skill design.} \tag{20}$$

$$\text{Employee lisa works on project p2 with skill modelling.} \tag{21}$$

$$\text{Employee lisa works on project p2 with skill planning.} \tag{22}$$

$$\text{Employee peter works on project p1 with skill coding.} \tag{23}$$

$$\text{Employee peter works on project p1 with skill planning.} \tag{24}$$

$$\text{Employee sarah works on project p1 with skill coding.} \tag{25}$$

$$\text{Employee sarah works on project p1 with skill modelling.} \tag{26}$$

$$\text{Employee sarah works on project p2 with skill modelling.} \tag{27}$$

$$\text{Employee sarah is project leader for project p1.} \tag{28}$$

$$\text{Employee lisa is project leader for project p2.} \tag{29}$$

Even though this list of sentences sounds artificial, it is still more human-readable than the answer set itself, especially if all predicates are shown and not only the output predicates. Without any filtering, the full answer set is usually very large since it incorporates the encoded problem instance as well as all auxiliary predicates in the ASP code (not shown here).

Note that before the sentences are generated, all instantiations of an atom are sorted alphabetically by the first term. This step is taken because the atoms in an answer set are in a random order by default.

"Contracting" the Sentences. Up to now we have not really employed the syntactic information provided by the parser for the atom descriptions. All we had used so far was the mapping between the variables in the sentences and the variables given by the @atom signatures.

Let us now consider sentences (14) to (18) of the above interpretation involving employee Boris. We will illustrate on them how one may make use of the syntactic information for "contracting" the sentences.

As is easily noticeable, these sentences offer redundant information. This property was to be expected since we only have one boilerplate sentence for each atom, which, however, usually has many instantiations in one answer set. Thus, placing these sentences without modification next to each other is most probably going to repeat already known information. This repetition is going to be especially severe if the number of instantiations of an atom is very large. Consequently, we have to look for procedures to systematically condense the information contained in the basic sentences so that they may expressed more concisely.

One Varying Element. The first approach is to focus on sentences which differ only in one variable. Consider sentences (14) and (15) on the one hand and (16) to (18) on the other hand. Within each respective group, the sentences differ only in the last variable, viz. concerning the variable denoting the skill which is assigned to an employee for a certain project. The best way to summarise the sentences of this kind of a group is to coordinate the varying elements, i.e., the differing skills, while maintaining the constituents that the sentences share, i.e., the employee and the project. In the first group, the varying constituents are with skill design and with skill marketing. Knowing that the constituents have the same PP shells, we can implement the rule that the NPs within the PPs can be coordinated, the noun phrase between the preposition and the term being pluralised if possible and only displayed in the first conjunct, and finally the preposition added in the front. This simple rule results in the contracted phrase with skills design and marketing. If we proceed as described, we are able to cut down the number of sentences as follows:

> Employee boris works on project p1 with skills design and marketing. (30)

> Employee boris works on project p2 with skills design, marketing, and planning. (31)

Note that whilst the noun *skill* is used in its singular form in the original sentences (14) to (18), here the plural form is employed. Recall, as pointed out in Sect. 3.2, the information what the plural form looks like and on which word it has to appear can be specified in NP constituents in the atom description. In instances where the varying elements are grouped into one coordination phrase, it is usually quite safe to use the plural form (if available).

In general, this contraction step can be performed whenever there is a group of sentences where there is one element (i.e., an instantiated variable) varying across the board, while all remaining elements stay the same in all sentences of the group. The preliminary grouping process is crucial for the output: depending on how the sentences are grouped, the varying element may change. In the example above, the default grouping was applied so that the varying variable was the last variable in the sentence. However, a different grouping may be possible, where the varying variable is, e.g., the first one in the sentence. In this case, we coordinate the subjects and end up with the verb morphologically agreeing with the coordinated subjects, i.e., it shows plural morphology. A possible sentence that would reflect this grouping is the following:

```
Employee hans and sarah work on project p1 with skill modelling.
```

Finally, if we take the complete answer set from Fig. 2 and apply the described contracting process (with the default grouping/sorting order), we get the following condensed version:

```
Employee boris works on project p1 with skills design and
marketing.
Employee boris works on project p2 with skills design,
marketing, and planning.
Employee hans works on project p1 with skill modelling.
Employee lisa works on project p2 with skills design,
modelling, and planning.
Employee peter works on project p1 with skills coding
and planning.
Employee sarah works on project p1 with skills coding
and modelling.
Employee sarah works on project p2 with skill modelling.

Employee sarah is project leader for project p1.
Employee lisa is project leader for project p2.
```

As we can see, taking this measure reduces the number of sentences of the original interpretation from 16 to 9.

One Identical Element. After the first contraction step, there may still be some redundancy left, though. In order to address this, we use the *topic-rheme dichotomy* (cf., e.g., Halliday [13]) to characterise the information structure of a certain kind of redundancy: While the topic ("what is being talked about") stays the same throughout a group of sentences, the rheme ("what is being said about the topic") changes in each sentence. Hence, in sentences (30) and (31) above, the topic is *boris* and the rheme is his job assignment on the two projects. In most English sentences the topic corresponds to the subject of the sentence.

In the described configuration, a second contraction step can be performed thus: The sentences in such a group can be coordinated, with the common topic functioning as a common subject for the coordination. Hence, if we condense sentences (30) and (31) accordingly, we obtain the following:

> Employee boris works on project p1 with skills design and
> marketing and on project p2 with skills design, marketing,
> and planning.

In this example we have taken the constituents that make up the varying rheme and coordinated them. The coordinated phrases can then be added to the fixed topic and the associated verb. By applying this contraction to the whole answer set, we can lower the number of sentences to 7.

By using the structural information provided by the grammar and ultimately by the parse tree, we are able to identify the borders of constituents easily, which allows us insert newly generated phrases at the appropriate spots. Also, without the syntactic tagging, we would not be able to distinguish constituents from non-constituents. This very basic identification is required in order to specify rules on similar constituents, i.e., constituent boilerplates where only the term varies.

5   The Eclipse Plug-in

We have implemented our tool as an Eclipse View, called "Lana Interpretation View", to be used in connection with the IDE SeaLion for ASP [2]. Using Eclipse's built-in repositioning feature, one can have both the Interpretation View of SeaLion and the Lana Interpretation View side-by-side, as shown in Fig. 3. This way, the user can choose an answer set in the Interpretation View and directly see the generated interpretation for it.

The Lana Interpretation View is divided into two main parts: the left part showing the instantiations of the atoms marked as output predicates, and the right part providing an editor with the generated textual interpretation, which can be edited and exported.

On the very left there is a list widget containing atoms, which gets refreshed according to the selection in SeaLion's Interpretation View. This widget displays all atoms which are listed in the @output annotation, together with the name of the block that they belong to (annotated with @block in the code). The table of terms next to the list changes its values according to the selection in the list. This table simply displays all instantiations of the selected atom, allowing the user to get an overview of the terms and sort them according to their needs.

As already pointed out previously, the sorting order may change the generated interpretation since it determines the grouping algorithm, which in turn influences the output of the "contraction" steps. A particular sorting order is given in Fig. 4.

As one can see, the values in the table are sorted by project first, then by skills. We have marked the groups with rectangles, in which one term varies while all other terms stay stable across the instantiated atom in the group. This is relevant for the "one varying element" approach as described in Sect. 4. Since the employee terms are the "least sorted" element and, thus, the one varying element, they will be coordinated to reduce the number of sentences.

The other half of the view shows a simple text editor that displays the generated textual interpretation and another text widget pointing to problems encountered by the parser. The editor allows the user to make ad-hoc changes to the text if desired. The buttons on the right implement the export function.

The checkbox **Condensed** indicates whether the second contraction step has been applied. Checking the box will lead to a recomputation and the result is shown in

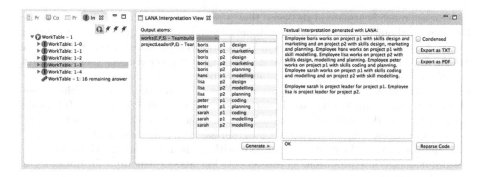

Fig. 3. SeaLion's Interpretation View and LANA Interpretation View next to each other.

Output atoms:

works(E,P,S) – Teambuild		
projectLeader(P,E) – Tear		
peter	p1	coding
sarah	p1	coding
boris	p1	design
boris	p1	marketing
hans	p1	modelling
sarah	p1	modelling
peter	p1	planning
boris	p2	design
lisa	p2	design
boris	p2	marketing
lisa	p2	modelling
sarah	p2	modelling
boris	p2	planning
lisa	p2	planning

Generate >

Fig. 4. The atom table with sorted values and with the "contraction groups" marked with rectangles.

the editor immediately afterwards. The **Reparse Code** button forces the system to reparse the code files that were specified for this corresponding program. This is useful when the feedback text widget indicates parsing errors in particular atom descriptions. The user is advised to inspect the sentences and check whether they form an acceptable sentence in our CNL. Once all corrections are made, a reparse should be performed.

6   Conclusion

In this paper, we presented a CNL approach for generating interpretations for answer sets in which the user can specify meta-information about the predicates used in an answer-set program. The descriptions in the LANA annotations are restricted according to the CNL we defined. Such an approach can primarily be useful during the development phase of an answer-set program when dealing with domain experts who are unfamiliar with logical methods. In such a setting, often small answer sets can already be helpful for detecting modelling errors. The question of dealing with suitable representations for large answer sets is challenging and an issue for future research.

Our approach is generic in the sense that it relies only on an answer set and meta-information about the atoms therein as input. Thus, in principle, it could be used also for other semantics and methods as well, like relational databases. Indeed, the problem setting we studied is similar to work in natural language processing for *generating* text out of structured data [14]. In the database community, visualisation techniques for representing relational data are important but such approaches are somewhat complementary to our goals. Likewise distinct to our work are approaches which provide *justifications* for the inclusion or non-inclusion of ground atoms in answer sets [17] or giving explanations why an answer-set program has no answer sets at all [25]. In fact, such approaches could be used in conjunction with methods like ours providing not only natural-language interpretations of answer sets but also explanations for obtaining a particular output.

Another interesting issue for future work is to investigate the question of translating answer-set programs *themselves* into natural language. However, this task requires a more dedicated syntactic and semantic analysis of user-specified sentences.

References

1. Baral, C.: Knowledge Representation, Reasoning and Declarative Problem Solving. Cambridge University Press, Cambridge (2003)
2. Busoniu, P., Oetsch, J., Pührer, J., Skocovsky, P., Tompits, H.: Sealion: an eclipse-based IDE for answer-set programming with advanced debugging support. Theory Pract. Log. Program. **13**(4–5), 657–673 (2013)
3. De Vos, M., Kiza, D., Oetsch, J., Pührer, J., Tompits, H.: Annotating answer-set programs in Lana. Theory Pract. Log. Program. **12**(4–5), 619–637 (2012)
4. Erdem, E., Erdogan, H., Öztok, U.: BioQuery-ASP: querying biomedical ontologies using answer set programming. In: Proceedings of 5th International RuleML2011@BRF Challenge, CEUR Workshop Proceedings, vol. 799. CEUR-WS.org (2011)
5. Erdem, E., Öztok, U.: Generating explanations for biomedical queries. Theory Pract. Log. Program. **15**(1), 35–78 (2015)
6. Erdem, E., Yeniterzi, R.: Transforming controlled natural language biomedical queries into answer set programs. In: Proceedings of Workshop on Current Trends in Biomedical Natural Language Processing, pp. 117–124 (2009)
7. Fang, M.: A controlled natural language approach for interpreting answer sets. B.Sc. thesis, Technische Universität Wien, Institute for Information Systems (2013)
8. Febbraro, O., Reale, K., Ricca, F.: ASPIDE: integrated development environment for answer set programming. In: Delgrande, J.P., Faber, W. (eds.) LPNMR 2011. LNCS (LNAI), vol. 6645, pp. 317–330. Springer, Heidelberg (2011). https://doi.org/10.1007/978-3-642-20895-9_37
9. Fuchs, N.E., Kaljurand, K., Kuhn, T.: Attempto controlled english for knowledge representation. In: Baroglio, C., Bonatti, P.A., Małuszyński, J., Marchiori, M., Polleres, A., Schaffert, S. (eds.) Reasoning Web. LNCS, vol. 5224, pp. 104–124. Springer, Heidelberg (2008). https://doi.org/10.1007/978-3-540-85658-0_3
10. Fuchs, N.E., Schwertel, U., Schwitter, R.: Attempto controlled English—not just another logic specification language. In: Flener, P. (ed.) LOPSTR 1998. LNCS, vol. 1559, pp. 1–20. Springer, Heidelberg (1999). https://doi.org/10.1007/3-540-48958-4_1

11. Gelfond, M., Lifschitz, V.: The stable model semantics for logic programming. In: Proceedings of 5th International Conference and Symposium on Logic Programming (ICLP/SLP 1988), vol. 88, pp. 1070–1080 (1988)

12. Guy, S., Schwitter, R.: The PENG ASP system: architecture, language and authoring tool. Lang. Resour. Eval. **51**(1), 67–92 (2017)

13. Halliday, M.A., Matthiessen, C.M.: An Introduction to Functional Grammar. Arnold Publishers, London (2004)

14. Indurkhya, N., Damerau, F.J.: Handbook of Natural Language Processing. CRC Press, Boca Raton (2010)

15. Kuhn, T.: A survey and classification of controlled natural languages. Comput. Linguist. **40**(1), 121–170 (2014)

16. Leone, N., et al.: The DLV system for knowledge representation and reasoning. ACM Trans. Comput. Log. **7**(3), 499–562 (2006)

17. Pontelli, E., Son, T.C., El-Khatib, O.: Justifications for logic programs under answer set semantics. Theory Pract. Log. Program. **9**(1), 1–56 (2009)

18. Potassco—The Potsdam Answer Set Solving Collection. http://potassco.sourceforge.net

19. Schwitter, R.: English as a formal specification language. In: Proceedings of 13th International Conference on Database and Expert Systems Applications (DEXA 2002), pp. 228–232. IEEE (2002)

20. Schwitter, R.: Working for two: a bidirectional grammar for a controlled natural language. In: Wobcke, W., Zhang, M. (eds.) AI 2008. LNCS (LNAI), vol. 5360, pp. 168–179. Springer, Heidelberg (2008). https://doi.org/10.1007/978-3-540-89378-3_17

21. Schwitter, R.: Controlled natural languages for knowledge representation. In: Proceedings of 23rd International Conference on Computational Linguistics (COLING 2010), pp. 1113–1121 (2010)

22. Schwitter, R.: Answer set programming via controlled natural language processing. In: Kuhn, T., Fuchs, N.E. (eds.) CNL 2012. LNCS (LNAI), vol. 7427, pp. 26–43. Springer, Heidelberg (2012). https://doi.org/10.1007/978-3-642-32612-7_3

23. Schwitter, R.: The jobs puzzle: taking on the challenge via controlled natural language processing. Theory Pract. Log. Program. **13**(4–5), 487–501 (2013)

24. Schwitter, R., Tilbrook, M.: Dynamic semantics at work. In: Sakurai, A., Hasida, K., Nitta, K. (eds.) JSAI 2003-2004. LNCS (LNAI), vol. 3609, pp. 416–426. Springer, Heidelberg (2007). https://doi.org/10.1007/978-3-540-71009-7_39

25. Syrjänen, T.: Debugging inconsistent answer set programs. In: Proceedings of 11th International Workshop on Non-Monotonic Reasoning (NMR 2006), pp. 77–83. Institut für Informatik, Technische Universität Clausthal, Technical report (2006)

26. Van Valin, R.D.: An Introduction to Syntax. Cambridge University Press, Cambridge (2001)

Techniques for Efficient Lazy-Grounding ASP Solving

Lorenz Leutgeb[1] and Antonius Weinzierl[1,2]

[1] Knowledge-Based Systems Group, Institute of Information Systems,
TU Wien, Vienna, Austria
`lorenz@leutgeb.xyz`, `weinzierl@kr.tuwien.ac.at`
[2] Department of Computer Science, Aalto University, Espoo, Finland

Abstract. Answer-Set Programming (ASP) is a well-known and expressive logic programming paradigm based on efficient solvers. State-of-the-art ASP solvers require the ASP program to be variable-free, they thus ground the program upfront at the cost of a potential exponential explosion of the space required. Lazy-grounding, where solving and grounding are interleaved, circumvents this grounding bottleneck, but the resulting solvers lack many important search techniques and optimizations. The recently introduced ASP solver Alpha combines lazy-grounding with conflict-driven nogood learning (CDNL), a core technique of efficient ASP solving. This work presents how techniques for efficient propagation can be lifted to the lazy-grounding setting. The Alpha solver and its components are presented and detailed benchmarks comparing Alpha to other ASP solvers demonstrate the feasibility of this approach.

1 Introduction

Answer-Set Programming (ASP) is an expressive logic programming paradigm where non-monotonic rules are used to formalize problem descriptions. The semantics of such rules are given in terms of answer sets, which represent solutions to the specified problem (see [4] for a detailed introduction). For example the following rules encode that for each node N of a graph a color C can be chosen or not be chosen.

$$chosenColor(N, C) \leftarrow node(N), color(C), \textbf{not } notChosenColor(N, C).$$
$$notChosenColor(N, C) \leftarrow node(N), color(C), \textbf{not } chosenColor(N, C).$$

Rules allow to easily encode complex problems like graph coloring. Finding the answers to such a problem, however, is hard and requires advanced techniques.

ASP solvers are traditionally based on a two-phase computation. First, the variables are removed from the input program by grounding and second, the

---

This work has been supported by the Austrian Science Fund (FWF) project P27730 and the Academy of Finland, project 251170.

D. Seipel et al. (Eds.): DECLARE 2017, LNAI 10997, pp. 132–148, 2018.
https://doi.org/10.1007/978-3-030-00801-7_9

ground program is solved by highly optimized algorithms for propositional problems. Prominent such ground-and-solve systems are DLV [11] and Clingo [5]. Unfortunately, the ground program is in the worst case exponential in the size of the non-ground program. This makes many real-world programs too big to fit in memory and is therefore referred to as the *grounding bottleneck* of ASP.

Lazy-grounding on the other hand interleaves the grounding and solving phases and thus overcomes the grounding bottleneck (cf. GASP [13], AsPeRiX [10], and Omiga [3]). Due to this interleaving, such solvers explore the (exponential) search space differently from CDNL-based solvers, making them very inefficient at solving problems that are trivial for ground-and-solve ASP systems. The Lazy-MX system [2] for the language of $FO(ID)$ follows a different approach and achieves lazy-grounding with efficient solving, but it is restricted to (some) subclass of ASP and requires manual translation.

The recently introduced ASP solver Alpha [15] combines CDNL-based search procedures with lazy-grounding to get the best of both worlds: fast search space exploration and avoidance of the grounding bottleneck at the same time.

Example 1. Consider the following program which selects from a domain at most one element:

$$dom(1). \quad \ldots \quad dom(12). \qquad sel(X) \leftarrow dom(X), \textbf{not } nsel(X).$$
$$\leftarrow sel(Y), sel(X), X \neq Y. \qquad nsel(X) \leftarrow dom(X), \textbf{not } sel(X).$$

Adding to this program one rule that forms a large cross-product over selected elements is enough to exhibit the grounding bottleneck.

$$p(X_1, X_2, X_3, X_4, X_5, X_6) \leftarrow sel(X_1), sel(X_2), sel(X_3), sel(X_4), sel(X_5), sel(X_6).$$

For solvers like Clingo, the amount of required memory increases dramatically when domain elements are added to *dom*. A domain size of 20 already requires several gigabytes of memory to ground, while the same program can be solved by lazy-grounding almost immediately and without such memory consumptions.

Blending lazy-grounding and CDNL solving is challenging for a number of reasons. First, usual CDNL solvers guess truth assignments for atoms while lazy-grounding solvers guess whether rules satisfying certain conditions fire or not. Second, atoms may only become *true* due to a rule that fires and must not become *true* due to constraints, since, e.g., the constraint $\leftarrow \textbf{not } a.$ is no justification for a being *true*. Therefore unit-propagation on nogoods may not simply set atoms to *true*. Introducing *must-be-true* as a third truth value fixes this problem, but requires intricate adaptions on the data structures for unit-propagation. Specifically, the two-watched-literals schema for nogoods is no longer sufficient due to it functioning only with two truth values. A solution to the first challenge is described in [15] (including details why a translation to two-valued search faces severe unresolved problems). Here, we provide first an overview to that solution and are otherwise concerned with the second challenge.

The contributions (after preliminary Sect. 2) of this work are as follows:

- presenting the novel architecture of the Alpha ASP solver (in Sect. 3) followed by an overview of the Alpha approach for blending lazy-grounding and CDNL-based search,
- an enhancement of the two-watched literals schema to obtain efficient propagation performance in the presence of a third truth value and nogoods that are extended with heads (in Sect. 4), and
- benchmarks of the resulting ASP solver Alpha (in Sect. 5), showing impressive improvements but also directions for future work (in Sect. 6).

2 Preliminaries

Let $\mathcal{C}$ be a finite set of constants, $\mathcal{V}$ be a set of variables, and $\mathcal{P}$ be a finite set of predicates with associated arities, i.e., elements of $\mathcal{P}$ are of the form p/k where p is the predicate name and k its arity. We assume each predicate name occurs with exactly one arity. The set $\mathcal{A}$ of (non-ground) atoms is then given by $\{p(t_1, \ldots, t_n) \mid p/n \in \mathcal{P}, t_1, \ldots, t_n \in \mathcal{C} \cup \mathcal{V}\}$. An atom $at \in \mathcal{A}$ is ground if no variable occurs in it; the set of variables occurring in at is denoted by $vars(at)$. The set of all ground atoms is denoted by $\mathcal{A}_{grd}$. A (normal) rule is of the form:

$$at_0 \leftarrow at_1, \ldots, at_k, \textbf{not } at_{k+1}, \ldots, \textbf{not } at_n.$$

where each $at_i \in \mathcal{A}$ is an atom, for $0 \leq i \leq n$. For such a rule r the head, positive body, negative body, and body are defined as $H(r) = \{at_0\}$, $B^+(r) = \{at_1, \ldots, at_k\}$, $B^-(r) = \{at_{k+1}, \ldots, at_n\}$, and $B(r) = \{at_1, \ldots, at_n\}$, respectively. A rule r is a constraint if $H(r) = \emptyset$, a fact if $B(r) = \emptyset$, and ground if each $at \in B(r) \cup H(r)$ is ground. The variables occurring in r are given by $vars(r) = \bigcup_{at \in H(r) \cup B(r)} vars(at)$. A literal l is positive if $l \in \mathcal{A}$, otherwise it is negative. A rule r is *safe* if all variables occurring in r also occur in its positive body, i.e., $vars(r) \subseteq \bigcup_{a \in B^+(r)} vars(a)$.

A program P is a finite set of safe rules. P is ground if each $r \in P$ is. A (Herbrand) interpretation I is a subset of the Herbrand base w.r.t. P, i.e., $I \subseteq \mathcal{A}_{grd}$. An interpretation I satisfies a literal l, denoted $I \models l$ if $l \in I$ for positive l and $l \notin I$ for negative l. I satisfies a ground rule r, denoted $I \models r$ if $B^+(r) \subseteq I \wedge B^-(r) \cap I = \emptyset$ implies $H(r) \subseteq I$ and $H(r) \neq \emptyset$. Given an interpretation I and a ground program P, the FLP-reduct P^I of P w.r.t. I is the set of rules $r \in P$ whose body is satisfied by I, i.e., $P^I = \{r \in P \mid B^+(r) \subseteq I \wedge B^-(r) \cap I = \emptyset\}$. I is an *answer set* of a ground program P if I is the subset-minimal model of P^I; the set of all answer sets of P is denoted by $AS(P)$.

A substitution $\sigma : \mathcal{V} \rightarrow \mathcal{C}$ is a mapping of variables to constants. Given an atom at the result of applying a substitution σ to at is denoted by $at\sigma$; this is extended in the usual way to rules r, i.e., $r\sigma$ for a rule of the above form is $at_0\sigma \leftarrow at_1\sigma, \ldots, \textbf{not } at_n\sigma$. Then, the grounding of a rule is given by $grd(r) = \{r\sigma \mid \sigma \text{ is a substitution for all } v \in vars(r)\}$ and the grounding $grd(P)$ of a program P is given by $grd(P) = \bigcup_{r \in P} grd(r)$. Elements of $grd(P)$ and

$grd(r)$ are called ground instances of P and r, respectively. The answer sets of a non-ground program P are given by the answer sets of $grd(P)$.

CDNL-based ASP solving takes a ground program, translates it into nogoods and then runs a SAT-inspired (i.e., a DPLL-style) model building algorithm to find a solution for the set of nogoods. Following established notation, a Boolean signed literal is of the form $\mathbf{T}at$ and $\mathbf{F}at$ for $at \in \mathcal{A}$. A nogood $ng = \{s_1, \ldots, s_n\}$ is a set of Boolean signed literals s_i, $1 \leq i \leq n$, which intuitively states that a solution cannot satisfy all literals s_1 to s_n. For example, the nogood $ng = \{\mathbf{T}a, \mathbf{F}b\}$ states that it cannot be the case that a is *true* and b is *false* at the same time. Nogoods are interpreted over assignments, which are sets A of Boolean signed literals, i.e., an assignment is a (partial) interpretation where *false* atoms are represented explicitly. A solution for a set Δ of nogoods then is an assignment A such that $\{at \mid \mathbf{T}at \in A\} \cap \{at \mid \mathbf{F}at \in A\} = \emptyset$, $\{at \mid \mathbf{T}at \in A\} \cup \{at \mid \mathbf{F}at \in A\} = \mathcal{A}$, and no nogood is violated, i.e., $\forall ng \in \Delta : ng \not\subseteq A$. A solution thus corresponds one-to-one to an interpretation that is a model of all nogoods. For more details and algorithms, see [5–7]. The complement of a Boolean signed literal s, denoted $\overline{s}$, is $\overline{\mathbf{T}a} = \mathbf{F}a$ and $\overline{\mathbf{F}a} = \mathbf{T}a$. Also note that CDNL-based solvers for ASP employ additional checks to ensure that the constructed model is supported and unfounded-free, but these checks are not necessary in the approach presented.

Lazy-grounding, also called grounding on-the-fly, is built on the idea of a computation, which is a sequence $(A_0, \ldots, A_\infty)$ of assignments starting with the empty set and adding at each step heads of applicable rules (cf. [3,8,13]). A ground rule r is *applicable* in a step A_k, if its positive body already has been derived and its negative body is not contradicted, i.e., $B^+(r) \subseteq A_k$ and $B^-(r) \cap A_k = \emptyset$. Observe that finding applicable ground rules, i.e., finding a non-ground rule r and a grounding substitution σ such that $r\sigma$ is applicable, is the task of the (lazy) grounder. A computation $(A_0, \ldots, A_\infty)$ then has to satisfy the following conditions besides $A_0 = \emptyset$, given the usual immediate-consequences operator T_P:

1. $\forall i \geq 1 : A_i \subseteq T_P(A_{i-1})$ (the computation contains only consequences),
2. $\forall i \geq 1 : A_{i-1} \subseteq A_i$ (the computation is monotonic),
3. $A_\infty = \bigcup_{i=0}^{\infty} A_i = T_P(A_\infty)$ (the computation converges), and
4. $\forall i \geq 1 : \forall at \in A_i \setminus A_{i-1}, \exists r \in P$ such that $H(r) = at$ and $\forall j \geq i - 1 : B^+(r) \subseteq A_j \wedge B^-(r) \cap A_j = \emptyset$ (applicability of rules is persistent through the computation).

It has been shown that A is an *answer set* of a normal logic program P iff there is a computation $(A_0, \ldots, A_\infty)$ for P such that $A = A_\infty$ [9,12]. Observe that A is finite, i.e., $A_\infty = A_n$ for some $n \in \mathbb{N}$, because $\mathcal{C}, \mathcal{P}$, and P are finite.

3   The Alpha Approach

Architecture. Alpha is a combination of lazy-grounding and CDNL-search to obtain an ASP solver that avoids the grounding bottleneck and shows good search performance. On an abstract level, Alpha achieves this by utilizing a

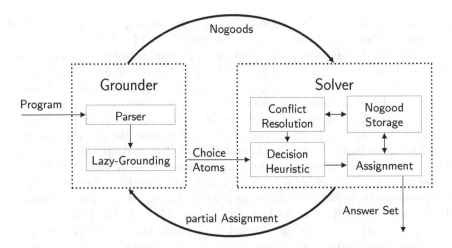

Fig. 1. Architecture of the Alpha system. Data flow is indicated by arrows. Grounder (left) and CDNL-based solver (right) interact cyclically for lazy-grounding.

grounder component and a solver component, where the solver is a modified CDNL-search algorithm, but both components interact cyclically in the style of lazy-grounding ASP systems. The architecture of the Alpha solver is depicted in Fig. 1. The grounder is composed of a parser and a semi-naive grounder that, given a partial assignment, computes all ground rules that potentially fire, transforms the ground rules into nogoods, and sends these to the solver. The solving component is a modified CDNL solver trying to find a satisfying assignment to the set of nogoods presented. It contains a nogood store for unit-propagation on nogoods, conflict resolution implementing conflict-driven nogood learning following the first-UIP schema to learn new nogoods, and a decision heuristic. The most important difference to an ordinary CDNL solver is that guessing is restricted to atoms representing applicable ground rules, i.e., rules whose positive body is satisfied in the current assignment. By that, Alpha prevents unfounded sets from becoming *true*, thus the assignments constructed by the solver are guaranteed to be unfounded-free. Another difference is that the partial assignments of Alpha contain truth values *true*, *false*, and *must-be-true* where the latter indicates that an atom must be true (e.g. due to a constraint) but no firing rule derives/justifies the atom yet.

Core Algorithm. The remainder of this section provides a summary of the Alpha algorithm and its fundamentals while full details can be found in [15]. The Alpha algorithm at a glance is given by Algorithm 1 which is similar to the main algorithm of CDNL solvers. There is one loop in which the search space is explored and each iteration begins with propagation from the already derived knowledge. If a conflict occurs, it is analyzed in line (a) and a new nogood is derived following the first-UIP schema for conflict-driven learning. In (b) the grounder is requested to derive new nogoods from the assignment derived so far. This is the

---

Algorithm 1: The Alpha Algorithm (simplified).

Input: A (non-ground) program P.
Output: The answer sets $\mathcal{AS}(P)$ of P.

Initialize $\mathcal{AS} = \emptyset$, assignment A, and nogood storage Δ.
Run lazy grounder, obtain initial nogoods Δ from facts.
while *search space not exhausted* **do**
   Propagate on Δ extending A.
   if *there exists conflicting nogood* **then**
     | Analyze conflict, learn new nogood, and backjump.   (a)
   else if *propagation extended A* **then**
     | Run lazy grounder w.r.t. A and extend Δ.   (b)
   else if *exists an applicable rule* **then**   (c)
     | Guess as chosen by heuristic.
   else if *exists unassigned atom* **then**   (d)
     | Assign all unassigned atoms to false.
   else if *no atom in A assigned to must-be-true* **then**   (e)
     | $\mathcal{AS} \leftarrow \mathcal{AS} \cup \{A\}$
     | Add enumeration nogood and backtrack.
   else   (f)
     | Backtrack.

return $\mathcal{AS}$

---

lazy-grounding part and it is usually absent in CDNL solvers. In (c) a heuristic decides which atom to guess on. This way of guessing has been newly developed for Alpha and it ensures that the atom guessed on corresponds exactly to an applicable ground rule, i.e., the positive body of the ground rule is already in the assignment and the negative body is not (yet) contradicted by the assignment. When (d) is reached, the interplay of propagation, grounding, and guessing has reached a fixpoint: there are no more applicable ground rule instances and nothing can be derived by propagation or from further grounding. However, there may still be some atoms with unassigned truth value, because the guessing is restricted and does not guess on all atoms. Therefore in (d) all unassigned atoms are explicitly assigned to *false* and the propagation at the beginning of the following iteration ensures that no nogood is violated. Finally, in (e) the solver checks whether there is an atom assigned to *must-be-true* but could not be derived by some rule firing to become *true*. If there is no *must-be-true*, the current assignment corresponds to an answer set and it is recorded as such. If the check fails, the current assignment is no answer set and backtracking occurs in (f).

In order to represent rules using nogoods, Alpha introduces the notion of a *nogood with head*, that is, a nogood $ng = \{s_1, \ldots, s_n\}_i$ with one distinguished negative literal s_i, $1 \leq i \leq n$, such that $s_i = \mathbf{F}a$ for some $a \in \mathcal{A}$. The head of a nogood is denoted by $hd(ng) = s_i$. The head literal, intuitively, captures the idea of the head of a logic programming rule: if the nogood is unit on the head, it is assigned to *true* and not just *must-be-true*.

The full representation of a rule by nogoods is as follows: let r be a rule and σ be a substitution such that $r\sigma$ is ground, let the positive body be $B^+(r\sigma) = \{a_1, \ldots, a_k\}$ and the negative body be $B^-(r\sigma) = \{a_{k+1}, \ldots, a_n\}$ while the head is $H(r\sigma) = \{a_0\}$, then the *nogood representation* is given by the following nogoods:

$$\{\mathbf{F}\beta(r,\sigma), \mathbf{T}a_1, \ldots, \mathbf{T}a_k, \mathbf{F}a_{k+1}, \ldots, \mathbf{F}a_n\}_1, \{\mathbf{F}a_0, \mathbf{T}\beta(r,\sigma)\}_1,$$
$$\{\mathbf{T}\beta(r,\sigma), \mathbf{F}a_1\}, \ldots, \{\mathbf{T}\beta(r,\sigma), \mathbf{F}a_k\}, \{\mathbf{T}\beta(r,\sigma), \mathbf{T}a_{k+1}\}, \ldots, \{\mathbf{T}\beta(r,\sigma), \mathbf{T}a_n\}$$

The new atom $\beta(r,\sigma)$, intuitively, represents the body of the ground rule $r\sigma$. Notice that the first and second nogood each has a head (as indicated by the subscript 1, the head is the first literal). Despite similarities, this nogood representation differs from the one used by Clingo: first, Alpha uses nogoods with heads and second, there are no nogoods establishing support of ground atoms, because that would require full grounding.

Example 2. Consider from Example 1 the rule r as follows:

$$sel(X) \leftarrow dom(X), \mathbf{not}\ nsel(X).$$

From an assignment A where $dom(3)$ holds, i.e., $\mathbf{T}dom(3) \in A$, the grounder generates the substitution $\sigma : X \mapsto 3$ for r and it introduces the new atom $\beta(r,3)$ representing the body of the ground rule $r\sigma$. It then yields the following nogoods:

$n_1 : \{\mathbf{F}\beta(r,3), \mathbf{T}dom(3), \mathbf{F}nsel(3)\}_1$ $n_2 : \{\mathbf{T}\beta(r,3), \mathbf{F}dom(3)\}$

$n_3 : \{\mathbf{T}\beta(r,3), \mathbf{T}dom(3)\}$ $n_4 : \{\mathbf{F}sel(3), \mathbf{T}\beta(r,3)\}_1$

Nogoods n_1 to n_3 establish that $\beta(r,3)$ holds if and only if the body of the ground rule holds. Nogood n_4 ensures that the head atom is *true* whenever $\beta(r,3)$ holds. Observe that n_1 and n_4 have their first literal indicated as head, i.e., the solver will not set them to *must-be-true* but to *true* whenever the nogood is unit and all other positively occurring literals are *true*. This enables the nogoods to represent rules in the presence of two truth values, *must-be-true* and *true*.

4   Efficient Propagation: 3-Watched-Literals

This section provides details on efficient propagation realized in Alpha. Our approach extends the state-of-the-art propagation technique from SAT and CDNL-based ASP solving known as the two-watched literals (2WL) schema (cf. [1]). A direct use of 2WL in lazy-grounding ASP solving, however, is not possible due to such solvers using *must-be-true* as a third truth value requiring special treatment. Since *must-be-true* allows propagation to *true*, but no other truth value may be changed once it is assigned, this requires a different propagation mechanism than 2WL, which is designed for propagation to *true* and *false* only.

Formally, propagation is the task of identifying nogoods that are unit, i.e., nogoods violated except for one yet unassigned literal whose truth value then is set in order to avoid violating the nogood, and subsequently assigning this unassigned literal. In Alpha, a nogood with head may propagate to the truth value *true, false,* and *must-be-true* while a nogood without head may only propagate to *false* and *must-be-true.* Subsequently, there are two notions of being unit: weakly-unit and strongly-unit. Formally, an *assignment* A in Alpha is over truth values $\mathbf{T}$, $\mathbf{F}$, and $\mathbf{M}$; the *Boolean-projection* $A^{\mathcal{B}}$ maps $\mathbf{M}$ to $\mathbf{T}$, i.e., $A^{\mathcal{B}} = \{\mathbf{T}a \mid \mathbf{T}a \in A \text{ or } \mathbf{M}a \in A\} \cup \{\mathbf{F}a \mid \mathbf{F}a \in A\}$. Given a nogood $ng = \{s_1, \ldots, s_n\}$ and an assignment A: ng is *weakly-unit* under A for s if $ng \setminus A^{\mathcal{B}} = \{s\}$ and $\overline{s} \notin A^{\mathcal{B}}$; ng is *strongly-unit* under A for s if ng is a nogood with head, $ng \setminus A = \{s\}$, $s = hd(ng)$, and $\overline{s} \notin A$. By this definition a nogood with head is strongly-unit only if all its positively occurring literals are assigned to *true.* Also note that only a nogood with head can be strongly-unit and if a nogood is strongly-unit, it also is weakly-unit.

Propagation is the least fixpoint of the *immediate unit-propagation,* i.e., $propagate(A) = lfp\big(\Gamma_{\Delta}(A)\big)$ s.t. for a set Δ of nogoods and an assignment A:

$$\Gamma_{\Delta}(A) = A \cup \{\mathbf{T}a \mid \exists \delta \in \Delta, \delta \text{ is strongly-unit under } A \text{ for } s = \mathbf{F}a\}$$
$$\cup \{\mathbf{M}a \mid \exists \delta \in \Delta, \delta \text{ is weakly-unit under } A \text{ for } s = \mathbf{F}a\}$$
$$\cup \{\mathbf{F}a \mid \exists \delta \in \Delta, \delta \text{ is weakly-unit under } A \text{ for } s = \mathbf{T}a\}$$

In order to compute the propagation efficiently, we extend the concept of two-watched literals to our setting where nogoods may have a head literal and a nogood can be unit in two different ways. Two-watched literals, intuitively is based on the following observations: if a nogood δ contains more than two literals $s_1, s_2, s_3 \in \delta$ that are unassigned in some assignment A and one of these, say s_3, becomes assigned in $A' \supset A$, then δ is still not unit. Hence for as long as there are at least two unassigned literals, the nogood need not be checked for being unit. Therefore each nogood only requires two of its unassigned literals to be watched for being assigned in order to detect when the nogood is unit.

For our setting where a nogood may be weakly-unit or strongly-unit, intuitively, two-watched literals are required twice, 2WL for each type of being unit. Since the literal that will be propagated by a strongly-unit nogood always is the head literal of the nogood, it need not be watched explicitly. Therefore, three watches are sufficient. These watches are organized such that each atom is assigned one list per polarity and unit-type. Notice that each nogood requires only three watches but for each atom there are four types of watches.

Definition 1. *A* watch structure W *for an assignment A and a set of nogoods Δ is a mapping $W : \mathcal{A} \rightarrow \Delta^4$ of atoms to quadruples of lists (sets) of nogoods in Δ. For a watch structure W, each atom $a \in \mathcal{A}$ is associated a quadruple of lists*

$$W(a) = \big(watch^+(a), watch^-(a), watch_{\alpha}^+(a), watch_{\alpha}^-(a)\big).$$

The list $watch^+(a)$ (resp. $watch^-(a)$) contains all nogoods δ where a watch is on a positive literal $\mathbf{T}a \in \delta$ (res. negative literal $\mathbf{F}a \in \delta$) for detecting whether δ is *weakly-unit.* The list $watch_{\alpha}^+(a)$, resp. $watch_{\alpha}^-(a)$, contains all nogoods δ where a watch is on a positive literal $\mathbf{T}a \in \delta$, resp. negative literal $\mathbf{F}a \in \delta$, for detecting whether δ is *strongly-unit.*

A visualization of this data structure is given in Fig. 2.

For convenience, in the following we denote for a signed literal $s = \mathbf{X}a$ by $watch(s)$ the list $watch^+(a)$ if $\mathbf{X} \in \{\mathbf{T}, \mathbf{M}\}$ and $watch^-(a)$ otherwise. Similarly, $watch_\alpha(s)$ denotes $watch_\alpha^+(a)$ if a $\mathbf{X} \in \{\mathbf{T}, \mathbf{M}\}$ and $watch_\alpha^-(a)$ if $\mathbf{X} = \mathbf{F}$.

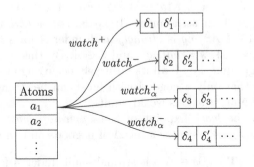

Fig. 2. Data structure for accessing watched nogoods.

In order to obtain correctly watched literals also after backtracking and subsequent assignments (where some assigned atoms may become unassigned and subsequently being propagated), the watches for satisfied nogoods have to point at those literals that were assigned in the highest decision level.

Given an assignment A and an atom a, we denote by $dl^w(A, a)$ the decision level on which a is assigned to *must-be-true* or *false* in A. Similarly, $dl^s(A, a)$ denotes the decision level at which a is assigned to *true* or *false* in A. Furthermore, for a signed literal $s = \mathbf{X}a$ with $\mathbf{X} \in \{\mathbf{F}, \mathbf{T}, \mathbf{M}\}$, we denote by $at(s)$ the atom of the literal, i.e., $at(s) = a$.

Intuitively, the watches of a nogood have to point at either (1) two unassigned literals, or (2) one of these literals atoms is assigned such that the nogood is satisfied and the other literal is either unassigned or assigned at an equal-or-higher decision level. The latter condition ensures that if backtracking removes the satisfying assignment then the second watched literal is guaranteed to be unassigned, i.e., even in case of backtracking the nogood is guaranteed to be either satisfied or contain two unassigned and watched literals.

Definition 2. *Let δ be a nogood and A be an assignment, then $s, s' \in \delta$ are potential watches if one of the following holds.*

(i) $at(s)$ *and* $at(s')$ *are both unassigned in A.*
(ii) The atom of s is complementary assigned, i.e., $\overline{s} \in A^{\mathcal{B}}$, and either s' is unassigned in A or $dl^w(A, at(s')) \geq dl^w(A, at(s))$.

For a nogood with head there is only one watch, which is not the head itself, and it obeys a similar condition; the main difference being that an atom assigned to *must-be-true* is treated like it were unassigned.

Definition 3. *Let δ be a nogood with head and A be an assignment, then $s_\alpha \in \delta$ with $hd(\delta) \neq s_\alpha$ is a potential α-watch if one of the following holds.*

(i) $at(s_\alpha)$ is unassigned in A, assigned to must-be-true in A, or s_α is complementary assigned in A.

(ii) $dl^s(A, at(s_\alpha)) \geq dl^s(A, at(hd(\delta)))$ and the head is true, i.e., $\mathbf{T}at(hd(\delta)) \in A$.

Example 3. Consider the assignment $A = \{\mathbf{M}c, \mathbf{F}d\}$ with $dl^w(A, c) \leq dl^w(A, d)$, i.e., $\mathbf{M}c$ was assigned at lower decision level than $\mathbf{F}d$, and the nogoods $\delta_1 = \{\mathbf{F}a, \mathbf{T}b, \mathbf{T}c, \mathbf{F}d, \mathbf{F}e\}_1$, $\delta_2 = \{\mathbf{F}c, \mathbf{F}d\}$, and $\delta_3 = \{\mathbf{F}a, \mathbf{T}c\}_1$, where δ_1 and δ_3 are nogoods with a head. For δ_1 any two literals from $\{\mathbf{F}a, \mathbf{T}b, \mathbf{F}e\}$ are potential watches since they are all unassigned and any literal in $\{\mathbf{T}b, \mathbf{T}c, \mathbf{F}e\}$ is a potential α-watch. The nogood δ_2 has the potential watches $\mathbf{F}c$ and $\mathbf{F}d$ since A assigns c complementary to its occurrence in δ_2 and d has higher decision level than c. Since δ_2 has no head, there is no potential α-watch. For δ_3 the literal $\mathbf{T}c$ is a potential α-watch since c is assigned *must-be-true* in A, but δ_3 has no potential watches, intuitively, because δ_3 is weakly-unit under A and propagates $\mathbf{F}a$.

Intuitively, a watch structure is consistent for an assignment and a set of nogoods, if each nogood is watched correctly.

Definition 4. *A watch structure W for a set of nogoods Δ is consistent with an assignment A if for each nogood $\delta \in \Delta$ there exist potential watches s, s' and, for δ being a nogood with head, a potential α-watch s_α such that $\delta \in watch(s)$, $\delta \in watch(s')$, and $\delta \in watch_\alpha(s_\alpha)$ all hold.*

Example 4 (continued). Let A be the same as in Example 3 and let $\Delta = \{\delta_1, \delta_2\}$. One watch structure W consistent with Δ and A is as follows:

$$W(a) = (\emptyset, \{\delta_1\}, \emptyset, \emptyset) \quad W(b) = (\{\delta_1\}, \emptyset, \{\delta_1\}, \emptyset) \quad W(c) = (\emptyset, \{\delta_2\}, \emptyset, \emptyset)$$
$$W(d) = (\emptyset, \{\delta_2\}, \emptyset, \emptyset) \quad W(e) = (\emptyset, \emptyset, \emptyset, \emptyset)$$

Thus W watches δ_1 on $\mathbf{F}a$, $\mathbf{T}b$, and α-watches it on $\mathbf{T}b$. Furthermore, it watches δ_2 on $\mathbf{F}c$ and $\mathbf{F}d$ while there exists no α-watch for δ_2 since it has no head. W is consistent because all watched literals in W are also potential (α-)watches in A. Note that for $\Delta' = \{\delta_1, \delta_2, \delta_3\}$ and A there exists no consistent watch structure since δ_3 has no potential watches (it is weakly-unit in A).

Computing *propagate*(A) is possible using Algorithm 2 where a watch structure W consistent with the current assignment A and set of nogoods Δ is maintained. Notice that the algorithm receives as input a set Σ of new assignments, i.e., assignments done by Algorithm 1 outside of propagation (for example by guessing or backtracking). Intuitively, Algorithm 2 iterates over all new assignments (including those it derives itself during propagation) until all new assignments have been processed. For each new assignment the two lists of watched nogoods fitting to the polarity of the assignment are considered, e.g., if $\mathbf{F}d$ is a new assignment then only nogoods δ with $\mathbf{F}d \in \delta$ are considered. Each of those lists is then checked whether one of its nogoods is violated, weakly-unit, or

---

Algorithm 2: *propagate*

Input: An assignment A, a set Σ of new assignments, and a watch structure W consistent with A and Δ.

Output: An (extended) assignment A' or a pair of extended assignment A' and a violated nogood d.

$A' \leftarrow A$

while $\Sigma \neq \emptyset$ **do**

$\quad\quad\Sigma \leftarrow \Sigma \setminus \{\mathbf{X}a\}$ for some $\mathbf{X}a \in \Sigma$.   // Process each new assignment.

$\quad\quad(\Delta, \Delta_\alpha) \leftarrow \begin{cases} (watch^+(a), watch^+_\alpha(a)) & \text{if } \mathbf{X} \in \{\mathbf{T}, \mathbf{M}\}, \text{ and} \\ (watch^-(a), watch^-_\alpha(a)) & \text{otherwise.} \end{cases}$

$\quad\quad$**foreach** $\delta \in \Delta$ **do**   // Propagation to $\mathbf{M}, \mathbf{F}$.

$\quad\quad\quad\quad$**if** δ *is violated* **then**

$\quad\quad\quad\quad\quad |$ **return** (A', δ)

$\quad\quad\quad\quad$**else if** δ *is weakly-unit for s* **then**

$\quad\quad\quad\quad\quad$ Let $s' = \mathbf{M}b$ if $s = \mathbf{F}b$ and $s' = \mathbf{F}b$ otherwise.

$\quad\quad\quad\quad\quad$ $A' \leftarrow A \cup \{s'\}$

$\quad\quad\quad\quad\quad$ $\Sigma \leftarrow \Sigma \cup \{s'\}$

$\quad\quad\quad\quad$Remove δ from Δ.   // Update ordinary watches.

$\quad\quad\quad\quad$Let s, s' be potential watches of δ

$\quad\quad\quad\quad$$watch(at(s)) \leftarrow watch(at(s)) \cup \{\delta\}$

$\quad\quad\quad\quad$$watch(at(s')) \leftarrow watch(at(s')) \cup \{\delta\}$

$\quad\quad$**foreach** $\delta \in \Delta_\alpha$ **do**   // Propagation to $\mathbf{T}$.

$\quad\quad\quad\quad$**if** δ *is strongly-unit* **then**

$\quad\quad\quad\quad\quad$ $A' \leftarrow A \cup \{\mathbf{T}at(hd(\delta))\}$

$\quad\quad\quad\quad\quad$ $\Sigma \leftarrow \Sigma \cup \{\mathbf{T}at(hd(\delta))\}$

$\quad\quad\quad\quad$Remove δ from Δ_α.   // Update alpha watch.

$\quad\quad\quad\quad$Let s be a potential α-watch of δ

$\quad\quad\quad\quad$$watch_\alpha(at(s)) \leftarrow watch_\alpha(at(s)) \cup \{\delta\}$

return A'

---

strongly-unit. If one of the latter two is the case, a new assignment is recorded. Afterwards, the watch structure is adapted such that consistency (with regard to the currently processed assignment) is restored. The following holds:

Proposition 1. *Let W be a watch structure W for a finite set of nogoods Δ that is consistent with an assignment A and let $A' \supseteq A$ be a larger assignment with $\Sigma = A' \setminus A$. Then, Algorithm 2 running on $A, \Sigma,$ and W returns either*

1. *a pair (A'', δ) such that A'' contains only consequences of A' and Δ and $\delta \in \Delta$ is violated by A'', or*
2. *an assignment $A'' = propagate(A')$ and the modified watch structure is consistent with A'' and Δ.*

Proof (sketch). First, observe that the outermost loop of Algorithm 2 terminates after finitely many iterations, because at each iteration one element of Σ is removed and there are only finitely many elements that can be added to Σ since Δ is finite and each $\delta \in \Delta$ contains only finitely many literals. Second, note that the assignment A' is extended by a signed literal $\mathbf{T}a, \mathbf{M}a$, or $\mathbf{F}a$ only if this literal is implied by a weakly- or strongly-unit nogood $\delta \in \Delta$. Hence every assigned literal in A' is a logical consequence of A' and Δ. Third, it holds that as long as no nogood is violated, every assignment s added to A' leads to all watches, that are no longer consistent with the extended assignment, becoming treated in subsequent iterations of the outermost loop, because whenever A' is extended so is Σ and the outermost loop stops only when $\Sigma = \emptyset$ or if some nogood is violated by A'. Next, we distinguish on the return value of Algorithm 2:

1. Algorithm 2 returns a pair (A'', δ). Then, A'' contains only consequences of A' and Δ by the second observation above and $\delta \in \Delta$ is violated by A'' as checked by the algorithm.
2. Algorithm 2 returns an assignment A''. Then, the outermost loop terminated due to $\Sigma = \emptyset$, i.e., every watch that was inconsistent with A'' and Δ has been replaced with a potential watch again. Furthermore, assignments to other than the watched literals have no influence on the watched ones continuing to be potential watches. Thus, the watch structure is consistent.

   It remains to show that $A'' = propagate(A')$. From the second observation above follows that $A'' \subseteq propagate(A')$, thus it only remains to show that every assignment $s \in propagate(A')$ is contained in A''. Towards contradiction, let $S \subseteq propagate(A')$ with $S \cap A'' = \emptyset$ be the largest set of assignments missing in A''. Pick $s \in S$ such that s is directly implied by unit-propagation on some $\delta_s \in \Delta$ and A''. Since $propagate(A') = lfp(\Gamma_\Delta(A'))$ induces a well-ordering on assignments, where the order is based on being directly implied by unit-propagation, it follows that such a "smallest" $s \in S$ and $\delta_s \in \Delta$ exist. Consequently, δ_s is unit under A'' for s. This directly contradicts that the watch structure W is consistent after Algorithm 2 finished. Hence, no such s exists, $S = \emptyset$, and therefore $A'' = propagate(A')$. $\qquad\square$

5 Evaluation

We evaluated the Alpha solver on four benchmarks, that exercise different parts of a solver, comparing Alpha to the lazy-grounding solvers Omiga [3] and AsPeRiX [10] as well as to Clingo [7]. All benchmarks were performed on a Linux machine with two 12-core AMD Opteron 6176 SE CPUs and 128 GB RAM. The timeout for each run was 300 s and the memory limit 8 GB. The *HTCondor* system (cf. http://research.cs.wisc.edu/htcondor) was used for load distribution to minimize runtime variations for different runs. Since Java restricts itself to use only parts of the available system memory, the JVM was instructed that 8 GB of RAM are available and that it can use up to 3.5 GB for heap allocations, i.e., Java was called with the following command-line arguments `-XX:MaxRAM=8000M -Xmx3500M`. We report the average runtimes in seconds on 10 randomly generated instances for each benchmark problem, except for one benchmark where only one instance per size exists. The compared solver versions

Table 1. Grounding explosion benchmark results. Instance size is the overall number of constants in the domain. Shown is runtime in seconds; out of memory by memout.

Instance size	Alpha	Omiga	AsPeRiX	Clingo
8	1.37(0)	0.42(0)	5.54(0)	1.74(0)
10	1.48(0)	0.43(0)	0.02(0)	7.00(0)
12	1.46(0)	0.44(0)	0.02(0)	22.47(0)
14	1.64(0)	0.47(0)	0.02(0)	56.39(0)
16	1.60(0)	0.51(0)	0.03(0)	145.28(0)
18	1.64(0)	0.45(0)	0.03(0)	Memout
20	1.83(0)	0.46(0)	0.05(0)	Memout
500	2.19(0)	1.41(0)	1.06(0)	Memout
1000	2.30(0)	1.66(0)	2.21(0)	Memout

were: Clingo version 5.2.0, AsPeRiX version 2.5, Omiga built from source using Git commit 037b3f9 and Alpha from source using Git commit a65421f.

Ground Explosion. This is the program of Example 1, i.e., given some domain, select exactly one element from the domain and derive a new atom containing the selected element six times. Table 1 shows the runtimes for domain sizes from 8 up to 1.000 where each solver is requested to compute 10 answer sets.

All lazy-grounding ASP solvers compute the answer sets within seconds for all instances, while Clingo runs out of 8 GB memory with a domain of size 18 already. Comparing Alpha with Omiga and AsPeRiX one can observe that Alpha is slower than the other two. This is likely due to Alpha maintaining the data structures of a CDNL solver while Omiga and AsPeRiX use a more direct representation of rules. One surprising result is that AsPeRiX takes more than 5 s for the instance with domain of size 8, which is much higher than for larger instances. A closer investigation revealed that AsPeRiX needs a lot of time to detect when no more answer sets exist, which only shows in this particular instance where there exist less than the requested 10 answer sets. Requesting 14 answer sets from AsPeRiX for the instance with domain size 12 already results in a timeout. Alpha, in contrast, does not exhibit the same problem.

Cutedge Benchmarks. This problem was first introduced in [3] and is as follows: given a graph $G = (V, E)$, choose one edge $e \in E$ and compute reachability on the graph G' where e is cut, i.e., $G' = (V, E \setminus \{e\})$. This problem is hard for ASP systems that are based on grounding the program upfront, while it is significantly easier for lazy-grounding ASP solvers. This problem was run on graphs with 100 to 500 vertices and 3.000 to 125.000 edges and the solvers instructed to compute 10 answer sets each. The results are given in Table 2. As expected, Clingo is only able to solve small instances and starting from graphs with 12.000 edges Clingo always hits the timeout. Surprisingly, Clingo hits the timeout and does not run out of memory within 300 s. Further testing showed

Table 2. Cutedge benchmark results. Instance is number of vertices/average percentage of edge being present. Shown is the average runtime in seconds over 10 instances with number of timeouts in parentheses.

Instance size	Alpha	Omiga	AsPeRiX	Clingo
100/30	12.59(0)	4.25(0)	0.78(0)	27.64(0)
100/50	11.87(0)	6.22(0)	1.79(0)	79.50(0)
200/30	22.90(0)	13.46(0)	13.29(0)	300.00(10)
200/50	45.95(0)	24.20(0)	35.18(0)	300.00(10)
300/10	16.92(0)	10.08(0)	8.54(0)	291.35(4)
300/30	59.58(0)	32.36(0)	72.09(0)	300.00(10)
500/10	62.46(0)	32.01(0)	70.38(0)	300.00(10)
500/30	300.00(10)	122.16(0)	300.00(10)	300.00(10)
500/50	300.00(10)	215.01(0)	300.00(10)	300.00(10)

Clingo running out of memory when given more time. Table 2 further shows that Alpha is comparable to AsPeRiX and both are slower than Omiga. This may be rooted in the fact that Omiga uses a Rete network for efficient grounding while Alpha uses a semi-naive grounding procedure similar to that of AsPeRiX.

Graph Colorability. This problem is inspired by the problem with the same name from the ASP competition. The task is to color a given graph with 5 available colors. This problem poses no grounding problem but requires efficient search procedures. The benchmark was run on randomly generated instances with 10 to 1.000 vertices and 40 to 4.000 edges between two randomly selected nodes, i.e., no further structure was imposed. For each setting 10 random graphs were constructed. The average runtimes in seconds is reported in Table 3.

As expected, this benchmark is very easy for Clingo, while the lazy-grounding solvers Omiga and AsPeRiX struggle for all but the trivial instances. AsPeRiX performs better than Omiga, solving instances with 100 vertices and 200 edges. Such graphs are very sparse, however, and nearly each coloring yields an answer set. For less-trivial instances with more edges per vertex, like those with 30 vertices and 120 edges, Omiga and AsPeRiX time out on nearly all of them. Alpha is able to solve also the harder instances where search is non-trivial. Comparing Alpha with Clingo we observe that there still is a significant gap in terms of search performance. This is rooted in the fact that Clingo employs numerous efficient search techniques (heuristics, nogood forgetting, nogood minimization, etc.) that are largely lacking in Alpha. There is some progress on implementing heuristics in Alpha (cf. [14]), but due to the specifics of lazy-grounding (restricted guessing, etc.) the techniques of Clingo cannot be adapted directly.

In order to more precisely compare the lazy-grounding solvers, Table 4 shows their runtimes on graphs with a fixed number of 50 vertices and an increasing number of edges. Omiga has timeouts even for 50 edges while AsPeRiX is able

Table 3. Graph 5-colorability benchmark results. Instance is number of vertices/number of edges. Shown is the average runtime in seconds over 10 instances with number of timeouts in parentheses.

Instance size	Alpha	Omiga	AsPeRiX	Clingo
10/40	1.41(0)	14.33(0)	31.10(1)	0.02(0)
20/80	1.53(0)	234.93(6)	128.79(4)	0.02(0)
30/120	1.59(0)	300.00(10)	230.23(7)	0.03(0)
40/160	2.54(0)	300.00(10)	217.17(7)	0.04(0)
50/200	2.31(0)	300.00(10)	300.00(10)	0.04(0)
100/400	4.24(0)	300.00(10)	300.00(10)	0.06(0)
400/1600	22.54(0)	300.00(10)	300.00(10)	0.45(0)
1000/4000	119.94(0)	300.00(10)	300.00(10)	2.66(0)

Table 4. Graph 5-colorability benchmark on graphs with 50 vertices varying edges. Instance is number of vertices/number of edges. Shown is the average runtime in seconds over 10 instances with number of timeouts in parentheses.

Instance size	Alpha	Omiga	AsPeRiX	Clingo
50/50	1.88(0)	290.47(9)	0.24(0)	0.03(0)
50/100	2.05(0)	300.00(10)	0.45(0)	0.03(0)
50/200	2.31(0)	300.00(10)	300.00(10)	0.04(0)
50/300	74.39(2)	300.00(10)	300.00(10)	0.07(0)
50/500	168.76(4)	300.00(10)	300.00(10)	0.04(0)

to handle 100 edges. With more than 100 edges Alpha is the only lazy-grounding solver that returned the requested answer sets in time.

Reachability. This benchmark is comprised of a simple positive program computing reachability in a large graph. The task is: given some start vertex of a graph, compute the set of all vertices reachable from the start vertex. The tests were run on 10 randomly generated graphs for each instance size, with 1.000 and 10.000 vertices and 4.000 to 80.000 edges. Since the resulting ASP program contains no negation, Clingo only uses its intelligent grounder while the solver has no work left to do. The benchmark therefore compares the speed (and overhead) of lazy-grounding with a highly-optimized grounder. Table 5 shows the results: on large instances, Alpha is the fastest of all lazy-grounding solvers while for smaller instances Omiga and AsPeRiX are faster. The optimizations for solving purely positive programs in Clingo make it the fastest here.

Summary. We observe that Alpha is comparable in speed to the other lazy-grounding solvers for problems where lazy-grounding avoids the grounding bottleneck. In addition to that, Alpha provides much better search performance,

Table 5. Reachability benchmark results. Instance size is number of vertices/multiple of edges of the random graph.

Instance size	Alpha	Omiga	AsPeRiX	Clingo
1000/4	2.13(0)	1.21(0)	0.77(0)	0.11(0)
1000/8	3.19(0)	1.63(0)	2.57(0)	0.21(0)
10000/2	10.95(0)	7.82(0)	31.11(0)	0.52(0)
10000/4	13.06(0)	22.55(0)	130.00(0)	1.09(0)
10000/8	16.62(0)	56.93(0)	300.00(10)	2.27(0)

making search-intense problems solvable using lazy-grounding. There are, however, many efficient solving techniques not yet available since each must be checked and adapted for not relying on the knowledge of all ground instances (e.g. for program simplification). Thus Alpha is slower than state-of-the-art ASP solvers on problems where grounding is not an issue. For ASP programs where grounding is problematic, however, Alpha is the best choice as it provides a good compromise between grounding performance and solving performance. Alpha is freely available at: https://github.com/alpha-asp/Alpha.

6   Conclusion

We presented the novel ASP solver Alpha which combines lazy-grounding and CDNL-search to obtain a system that is both, avoiding the grounding bottleneck and efficiently exploring the search space. An overview of Alpha and its architecture was given. To provide an efficient propagation the well-known two-watched literals schema was enhanced to 3-watched literals in order to cope with nogoods being unit in two distinct ways. Benchmarks show that Alpha is on-par with other lazy-grounding solvers on problems where grounding is an issue, while it provides a significant improvement for problems where search is dominating. Due to its recency, Alpha lacks several optimizations for search, making it noticeably slower than Clingo. Contrary to Clingo, however, Alpha does not suffer from the grounding bottleneck. Topics for future work are forgetting of learned nogoods, and using dependency information like strongly-connected-components.

References

1. Biere, A., Heule, M., van Maaren, H., Walsh, T. (eds.): Handbook of Satisfiability, Frontiers in Artificial Intelligence and Applications, vol. 185. IOS Press, Amsterdam (2009)
2. de Cat, B., Denecker, M., Bruynooghe, M., Stuckey, P.J.: Lazy model expansion: interleaving grounding with search. J. Artif. Intell. Res. **52**, 235–286 (2015)
3. Dao-Tran, M., Eiter, T., Fink, M., Weidinger, G., Weinzierl, A.: OMiGA: an open minded grounding on-the-fly answer set solver. In: del Cerro, L.F., Herzig, A., Mengin, J. (eds.) JELIA 2012. LNCS (LNAI), vol. 7519, pp. 480–483. Springer, Heidelberg (2012). https://doi.org/10.1007/978-3-642-33353-8_38

4. Eiter, T., Ianni, G., Krennwallner, T.: Answer set programming: a primer. In: Tessaris, S., et al. (eds.) Reasoning Web 2009. LNCS, vol. 5689, pp. 40–110. Springer, Heidelberg (2009). https://doi.org/10.1007/978-3-642-03754-2_2

5. Gebser, M., Kaufmann, B., Neumann, A., Schaub, T.: *clasp*: A conflict-driven answer set solver. In: Baral, C., Brewka, G., Schlipf, J. (eds.) LPNMR 2007. LNCS (LNAI), vol. 4483, pp. 260–265. Springer, Heidelberg (2007). https://doi.org/10.1007/978-3-540-72200-7_23

6. Gebser, M., Kaufmann, B., Neumann, A., Schaub, T.: Conflict-driven answer set enumeration. In: Baral, C., Brewka, G., Schlipf, J. (eds.) LPNMR 2007. LNCS (LNAI), vol. 4483, pp. 136–148. Springer, Heidelberg (2007). https://doi.org/10.1007/978-3-540-72200-7_13

7. Gebser, M., Kaufmann, B., Schaub, T.: Conflict-driven answer set solving: from theory to practice. Artif. Intell. **187**, 52–89 (2012)

8. Lefèvre, C., Beatrix, C., Stephan, I., Garcia, L.: ASPeRIX, a first-order forward chaining approach for answer set computing. In: TPLP, pp. 1–45, January 2017

9. Lefèvre, C., Nicolas, P.: A first order forward chaining approach for answer set computing. In: Erdem, E., Lin, F., Schaub, T. (eds.) LPNMR 2009. LNCS (LNAI), vol. 5753, pp. 196–208. Springer, Heidelberg (2009). https://doi.org/10.1007/978-3-642-04238-6_18

10. Lefèvre, C., Nicolas, P.: The first version of a new ASP solver. In: Erdem, E., Lin, F., Schaub, T. (eds.) LPNMR 2009. LNCS (LNAI), vol. 5753, pp. 522–527. Springer, Heidelberg (2009). https://doi.org/10.1007/978-3-642-04238-6_52

11. Leone, N., et al.: The DLV system for knowledge representation and reasoning. ACM Trans. Comput. Log. **7**, 499–562 (2002)

12. Liu, L., Pontelli, E., Son, T.C., Truszczyński, M.: Logic programs with abstract constraint atoms: the role of computations. In: Dahl, V., Niemelä, I. (eds.) ICLP 2007. LNCS, vol. 4670, pp. 286–301. Springer, Heidelberg (2007). https://doi.org/10.1007/978-3-540-74610-2_20

13. Palù, A.D., Dovier, A., Pontelli, E., Rossi, G.: GASP: answer set programming with lazy grounding. Fundam. Inform. **96**(3), 297–322 (2009)

14. Taupe, R., Weinzierl, A., Schenner, G.: Introducing heuristics for lazy-grounding ASP solving. In: PAoASP (2017, to appear)

15. Weinzierl, A.: Blending lazy-grounding and CDNL search for answer-set solving. In: Balduccini, M., Janhunen, T. (eds.) LPNMR 2017. LNCS (LNAI), vol. 10377, pp. 191–204. Springer, Cham (2017). https://doi.org/10.1007/978-3-319-61660-5_17

The Syllogistic Reasoning Task: Reasoning Principles and Heuristic Strategies in Modeling Human Clusters

Emmanuelle-Anna Dietz Saldanha[1]([✉]), Steffen Hölldobler[1,2], and Richard Mörbitz[1]

[1] International Center for Computational Logic, TU Dresden, Dresden, Germany
{dietz,sh}@iccl.tu-dresden.de, richard.moerbitz@tu-dresden.de
[2] North-Caucasus Federal University, Stavropol, Russian Federation

Abstract. It seems widely accepted that human reasoning cannot be modeled by means of classical logic. Psychological experiments have repeatedly shown that participants' answers systematically deviate from the classical logically correct answers. Recently, a new computational logic approach to modeling human syllogistic reasoning has been developed which seems to perform better than other state-of-the-art cognitive theories. We take this approach as starting point, yet instead of trying to model *the* human reasoner, we aim at identifying clusters of reasoners, which can be characterized by reasoning principles or by heuristic strategies.

1 Introduction

In recent years, a new cognitive theory based on the Weak Completion Semantics (WCS) has been developed. It has its roots in the ideas first expressed by Stenning and van Lambalgen [12], but is mathematically sound [5], and has been successfully applied to various human reasoning tasks. An overview can be found in [4]. Hence, it was natural to ask whether the WCS is competitive in syllogistic reasoning and how it performs with respect to the cognitive theories evaluated in the meta-analysis by Khemlani and Johnson-Laird [7]. Syllogisms are one of the oldest kinds of logical argument that date back to Aristotle. They are especially important in the field of Psychology, as they can be easily understood, yet they are sophisticated enough to require non-trivial reasoning. According to [7], an established theory for human syllogistic reasoning is a necessary step towards a unified cognitive theory of reasoning.

A syllogism consists of two premises and a conclusion. The syllogistic reasoning task is then: given the two premises, what conclusions are valid? Consider the following pair of syllogistic premises:

$$\textit{All a are b.} \qquad \textit{Some c are not b.} \qquad \text{(AO3)}$$

The premises can be interpreted as quantified statements. In first-order logic (FOL), *some c are not a* follows from these premises. However, according to [7],

© Springer Nature Switzerland AG 2018
D. Seipel et al. (Eds.): DECLARE 2017, LNAI 10997, pp. 149–165, 2018.
https://doi.org/10.1007/978-3-030-00801-7_10

Table 1. The moods and their formalization.   **Table 2.** The four figures.

Mood	FOL	Short		Premise 1	Premise 2
Affirmative universal	$\forall X(a(X) \rightarrow b(X))$	Aab	Figure 1	a-b	b-c
Affirmative existential	$\exists X(a(X) \wedge b(X))$	Iab	Figure 2	b-a	c-b
Negative universal	$\forall X(a(X) \rightarrow \neg b(X))$	Eab	Figure 3	a-b	c-b
Negative existential	$\exists X(a(X) \wedge \neg b(X))$	Oab	Figure 4	b-a	b-c

the majority of participants in experimental studies either concluded *some c are not a* or answered that *no valid conclusion* follows. Yet, these two responses exclude each other, i.e., it is unlikely that the participants who answered *no valid conclusion* are the same ones who answered *some c are not a*, and vice versa.

The possible quantifiers and figures of the premises are shown in Tables 1 and 2: Each premise can have one of four quantifiers called *moods*. The entities can appear in four different orders called *figures*. Thus we can abbreviate the example from above which consists of moods A and O and figure 3 with AO3.

In [8], cognitive principles under the WCS for modeling the logical form of the representation of quantified statements in human reasoning are identified. The approach achieved a match of 89% with respect to the conclusions given by the participants and based on the data reported in [7]. This result stands out, as the best of the twelve other state-of-the-art cognitive theories achieved only a match of 84%.

While reasoning with conditionals, humans seem to take certain assumptions for granted which, however, are not stated explicitly in the task description. As psychological experiments show, these assumptions seem not to be arbitrary but instead are systematic in the sense that they are repeatedly made by participants. Furthermore, some assumptions appear in various experiments, whereas other assumptions are only made in very few experiments or only by some participants. In order to identify and structure these assumptions, we view them as principles that are either applied or ignored by the participants who have to solve the task. As starting point, we take the syllogistic reasoning approach presented in [8]. However, a drawback of this approach is that only the matching with respect to the aggregated data is considered, i.e., the approach models *the* human reasoner. However, the above example and other examples such as cases of the Wason selection task reported in [9], serve as indication that *the* human reasoner does not exist, but instead we might better search for clusters of human reasoners. These clusters might be expressed by principles, i.e., some clusters might apply some principles that are not applied by other clusters. We also take into account the assumption that some humans do not reason at all to solve syllogistic reasoning tasks. We believe that they use heuristic strategies [13,14] and present a way to combine them within the WCS.

The paper is structured as follows: In Sect. 2 we present the principles for the representation of quantified statements, motivated by findings from cognitive science and philosophy of language. The WCS and the encoding of quantified statements within this approach are introduced in Sects. 3 and 4. In Sect. 5, the clusters and heuristics are discussed and an overall evaluation of the WCS is presented. In Sect. 6, we give an overview of our implementation of computing the conclusions that are drawn depending on the applied principles.

2   Principles About Quantified Statements

Eight principles for developing a logical form of quantified statements are presented. They originate from [1,8] except of the principles in Sects. 2.5 and 2.8.

2.1   Quantified Statements as Conditionals (conditionals)

Independent of the quantifiers mood, we formalize any relation between two objects of a quantified statement by means of a conditional such that the first object is the antecedent and the second object is the conclusion in the conditional. For instance, the statement *All a are b* is expressed as $\forall X(a(X) \rightarrow b(X))$.

2.2   Licenses for Inferences (licenses)

Given the quantified statement *all a are b*, a license for this inference can then be expressed by *all a that are not abnormal, are b* [12]. Given the previous formalization of this statement as $\forall X(a(X) \rightarrow b(X))$, we extend this conditional by conjoining $a(X)$ together with an abnormality predicate: $\forall X(a(X) \land \neg ab(X) \rightarrow b(X))$. Further, *nothing is abnormal with respect to X*, i.e., $\neg ab(X)$ is assumed.

2.3   Existential Import and Gricean Implicature (import)

Humans understand quantifiers differently due to a pragmatic understanding of the language. For instance, in natural language, humans normally do not quantify over things that do not exist. Consequently, *all a* implies *some a exists*. This appears to be in line with human reasoning and has been called the *Gricean implicature* [3]. It corresponds to what sometimes in literature is also called *existential import*.

2.4   Unknown Generalization (unknownGen)

Humans seem to distinguish between *some y are z* and *some z are y*, as the results reported by [7] show. Nevertheless, if we would represent *some y are z* by $\exists X(y(X) \land z(X))$ then this is semantically equivalent to $\exists X(z(X) \land y(X))$ because conjunction is commutative in FOL. Likewise, humans seem to distinguish between *some y are z* and *all y are z*. Accordingly, if we only observe that an object o belongs to y and z then we do not want to conclude both, *some y are z* and *all y are z*. In order to distinguish between *some y are z* and *all y are z*, we introduce the following principle: If we know that *some y are z*, then there must not only be an object o_1, which belongs to y and z but there must be another object o_2, which belongs to y and for which it is unknown whether it belongs to z.

2.5   Deliberate Generalization (deliberateGen)

If all of the principles introduced so far are applied to an existential premise, the only object about which an inference can be made is the one resulting from the existential import principle. This is because the abnormality introduced by the licenses for inferences principle and according to the unknown generalization principle has to be false for the object introduced by existential import, but it is unknown for other objects. There is, however, evidence that some humans still draw conclusions in such circumstances [7]. We believe that they do not take into account abnormalities regarding objects that are not related to the premise.

2.6   Converse Premise (converse)

Although there seems to be some evidence that humans distinguish between *some y are z* and *some z are y* (see the results reported in [7]) we propose that premises of the form Iab imply Iba and vice versa. If there is an object which belongs to y and z, then there is also an object which belongs to z and y. Similarly, we apply this principle for the E mood.

2.7   Search Alternative Conclusions to NVC (searchAlt)

Our hypothesis is that when participants are faced with a NVC conclusion (*no valid conclusion*), they might not want to accept this conclusion and proceed to check whether there exists unknown information that is relevant. This information may be explanations about the facts coming either from an existential import or from unknown generalization. We use only the first as source for observations, as they are used directly to infer new information.

2.8   Contraposition (contraposition)

In FOL, a conditional statement of the form $\forall(X)(a(X) \rightarrow b(X))$ is logically equivalent to its *contrapositive* $\forall(X)(\neg b(X) \rightarrow \neg a(X))$. This contraposition also holds for the syllogistic moods A and E. There is evidence in [7] that some of the participants make use of this equivalence when solving syllogistic reasoning tasks. We believe that when they encounter a premise with the mood A (e.g., *all a are b*), then they might reason with the contrapositive conditional as well.

3   Weak Completion Semantics

3.1   Contextual Logic Programs

Contextual logic programs are (data) logic programs extended by the truth-functional operator ctxt, called *context* [2]. *Contextual (logic) program clauses* are expressions of the forms $A \leftarrow L_1 \wedge \ldots \wedge L_m \wedge \mathsf{ctxt}(L_{m+1}) \wedge \ldots \wedge \mathsf{ctxt}(L_{m+p})$ (called *rules*), $A \leftarrow \top$ (called *facts*), $A \leftarrow \bot$ (called *negative assumptions*)[1] and $A \leftarrow \mathsf{U}$ (called *unknown assumptions*), where A is an atom and the L_i with $1 \leq i \leq m+p$ are literals. A is called *head* and $L_1 \wedge \ldots \wedge L_m \wedge \mathsf{ctxt}(L_{m+1}) \wedge \ldots \wedge \mathsf{ctxt}(L_{m+p})$ as well as $\top, \bot$ and U, standing for *true, false* and *unknown* respectively, are called *body* of the corresponding clauses. A *contextual program*, denoted by $\mathcal{P}$, is a finite set of contextual program clauses. $\mathsf{g}\mathcal{P}$ denotes the set of all ground instances of clauses occurring in $\mathcal{P}$. A is *defined in* $\mathsf{g}\mathcal{P}$ iff $\mathsf{g}\mathcal{P}$ contains a rule or a fact with head A. A is *undefined in* $\mathsf{g}\mathcal{P}$ iff A is not defined in $\mathsf{g}\mathcal{P}$. The set of all atoms that are undefined in $\mathsf{g}\mathcal{P}$ is denoted by $\mathsf{undef}(\mathcal{P})$. The *definition of A in* $\mathsf{g}\mathcal{P}$ is defined as $\mathsf{def}(A, \mathcal{P}) = \{A \leftarrow Body \mid A \leftarrow Body$ is a rule or a fact occurring in $\mathsf{g}\mathcal{P}\}$. $\neg A$ is *assumed in* $\mathsf{g}\mathcal{P}$ iff $\mathsf{g}\mathcal{P}$ contains a negative assumption with head A, $\mathsf{g}\mathcal{P}$ does not contain an unknown assumption with head A, and $\mathsf{def}(A, \mathcal{P}) = \emptyset$. We omit the word *contextual* when we refer to programs, if not stated otherwise.

3.2   Three-Valued Łukasiewicz Logic Extended by ctxt Connective

We consider the three-valued Łukasiewicz logic together with the ctxt connective, for which the corresponding truth values are $\top, \bot$ and U, meaning *true, false* and *unknown*, respectively. A *three-valued interpretation* I is a mapping from the set of logical formulas to the set of truth values $\{\top, \bot, \mathsf{U}\}$, represented as a pair $I = \langle I^\top, I^\bot \rangle$ of two disjoint sets of atoms: $I^\top = \{A \mid A$ is mapped to $\top$ under $I\}$ and $I^\bot = \{A \mid A$ is mapped to $\bot$ under $I\}$. Atoms which do not occur in $I^\top \cup I^\bot$ are mapped to U. The truth value of a given formula under I is determined according to the truth tables in Table 3. $I(F) = \top$ means that a formula F is mapped to true under I. A *three-valued model* $\mathcal{M}$ of $\mathcal{P}$ is a three-valued interpretation such that $\mathcal{M}(A \leftarrow Body) = \top$ for each $A \leftarrow Body \in \mathsf{g}\mathcal{P}$. Let $I = \langle I^\top, I^\bot \rangle$ and $J = \langle J^\top, J^\bot \rangle$ be two interpretations. $I \subseteq J$ iff $I^\top \subseteq J^\top$ and $I^\bot \subseteq J^\bot$. I is the *least model* of $\mathcal{P}$ iff for any other model J of $\mathcal{P}$ it holds that $I \subseteq J$.

3.3   Integrity Constraints

A set of *integrity constraints* $\mathcal{IC}$ consists of clauses of the form $\mathsf{U} \leftarrow Body$, where *Body* is a conjunction of literals and U denotes the unknown. Hence, an interpretation maps an integrity constraint to $\top$ iff *Body* is either mapped to U or $\bot$. Given an interpretation I and a set of integrity constraints $\mathcal{IC}$, I *satisfies* $\mathcal{IC}$ iff all clauses in $\mathcal{IC}$ are true under I.

---

[1] Under WCS, the negative assumption will become $A \leftrightarrow \bot$ and, hence, A has to be false. This can, however, be overwritten by other rules and facts (*defeating* the assumption).

Table 3. The truth tables for the connectives under the three-valued Łukasiewicz logic and for ctxt(L). L is a literal, $\top$, $\bot$, and U denote *true*, *false*, and *unknown*, respectively.

F	$\neg F$		$\wedge$	$\top$	U	$\bot$		$\vee$	$\top$	U	$\bot$		$\leftarrow$	$\top$	U	$\bot$		$\leftrightarrow$	$\top$	U	$\bot$		L	ctxt(L)
$\top$	$\bot$		$\top$	$\top$	U	$\bot$		$\top$	$\top$	$\top$	$\top$		$\top$	$\top$	$\top$	$\top$		$\top$	$\top$	U	$\bot$		$\top$	$\top$
$\bot$	$\top$		U	U	U	$\bot$		U	$\top$	U	U		U	U	$\top$	$\top$		U	U	$\top$	U		$\bot$	$\bot$
U	U		$\bot$	$\bot$	$\bot$	$\bot$		$\bot$	$\top$	U	$\bot$		$\bot$	$\bot$	U	$\top$		$\bot$	$\bot$	U	$\top$		U	$\bot$

3.4 Forward Reasoning: Least Fixed Point of $\Phi_\mathcal{P}$

For a given $\mathcal{P}$, consider the following transformation: 1. For each ground atom A which occurs as head of a clause in $g\mathcal{P}$, replace all clauses of the form $A \leftarrow Body_1, \ldots, A \leftarrow Body_m$ occurring in $g\mathcal{P}$ by $A \leftarrow Body_1 \vee \ldots \vee Body_m$. 2. Replace all occurrences of $\leftarrow$ by $\leftrightarrow$. The obtained ground set of equivalences is called the *weak completion* of $\mathcal{P}$ or $wc\mathcal{P}$. Consider the following semantic operator, which is due to Stenning and van Lambalgen [12]: Let $I = \langle I^\top, I^\bot \rangle$ be an interpretation. $\Phi_\mathcal{P}(I) = \langle J^\top, J^\bot \rangle$, where

$$J^\top = \{A \mid A \leftarrow Body \in \mathsf{def}(A, \mathcal{P}) \text{ and } Body \text{ is } true \text{ under } \langle I^\top, I^\bot \rangle\}$$
$$J^\bot = \{A \mid \mathsf{def}(A, \mathcal{P}) \neq \emptyset \text{ and}$$
$$Body \text{ is } false \text{ under } \langle I^\top, I^\bot \rangle \text{ for all } A \leftarrow Body \in \mathsf{def}(A, \mathcal{P})\}.$$

[5] showed that the weak completion of non-contextual programs always has a least model under Łukasiewicz logic, which can be obtained as the least fixed point of $\Phi_\mathcal{P}$. However, for programs with the ctxt operator this property only holds if the programs do not contain cycles [2]. In this paper, let $\mathcal{M}_\mathcal{P}$ denote the least fixed point of $\Phi_\mathcal{P}$. We define $\mathcal{P} \models_{wcs} F$ iff $\mathcal{M}_\mathcal{P}(F) = \top$.

3.5 Backward Reasoning: Explanations by Means of Abduction

An *abductive framework* $\langle \mathcal{P}, \mathcal{A}, \mathcal{IC}, \models_{wcs} \rangle$ consists of a program $\mathcal{P}$, a set $\mathcal{A}$ of abducibles, a set $\mathcal{IC}$ of integrity constraints, and the entailment relation $\models_{wcs}$. The set of abducibles is $\mathcal{A} = \{A \leftarrow \top \mid A \in \mathsf{undef}(\mathcal{P})\} \cup \{A \leftarrow \bot \mid A \in \mathsf{undef}(\mathcal{P}) \text{ and } \neg A \text{ is not assumed in } g\mathcal{P}\}$. Let $\langle \mathcal{P}, \mathcal{A}, \mathcal{IC}, \models_{wcs} \rangle$ be an abductive framework and the observation $\mathcal{O}$ a set of literals. $\mathcal{O}$ is *explainable* in $\langle \mathcal{P}, \mathcal{A}, \mathcal{IC}, \models_{wcs} \rangle$ iff there exists an $\mathcal{E} \subseteq \mathcal{A}$, such that $\mathcal{P} \cup \mathcal{E} \models L$ for all $L \in \mathcal{O}$ and $\mathcal{P} \cup \mathcal{E}$ satisfies $\mathcal{IC}$. $\mathcal{E}$ is then called *explanation* for $\mathcal{O}$ given $\mathcal{P}$ and $\mathcal{IC}$. We restrict $\mathcal{E}$ to be *minimal*, i.e. there does not exist any other explanation $\mathcal{E}' \subseteq \mathcal{A}$ for $\mathcal{O}$ such that $\mathcal{E}' \subseteq \mathcal{E}$.

Among the minimal explanations, it is possible that some of them entail a certain formula F while others do not. There exist two strategies to determine whether F is a valid conclusion in such cases. F follows *credulously*, if it is entailed by at least one explanation. F follows *skeptically*, if it is entailed by all explanations. Due to previous results on modeling human reasoning [4], skeptical abduction seems to be adequate.

Here, observations, are specified as $\mathcal{O}_{\mathcal{P}} = \{A \mid A \leftarrow \top \in \mathsf{def}(A, \mathcal{P})\}$. Usually, this set is further restricted by considering only facts that result from the application of certain principles. The idea is to find an explanation for each observation $A \in \mathcal{O}_{\mathcal{P}}$ after the fact $A \leftarrow \top$ has been removed from $\mathsf{g}\mathcal{P}$.

3.6   Encoding of Quantified Statements

Negation by Transformation (transformation). The logic programs we consider do not allow heads of clauses to be negative literals. A negative conclusion $\neg p(X)$ is represented by introducing an auxiliary formula $p'(X)$ together with the clause $p(X) \leftarrow \neg p'(X)$ and the integrity constraint $\mathsf{U} \leftarrow p(X) \wedge p'(X)$. This is a widely used technique in Logic Programming. Applying the principle licenses introduced in Sect. 2.2, the first clause is extended to $p(X) \leftarrow \neg p'(X) \wedge \neg ab_{npp}(X)$ and the assumption $ab_{npp}(X) \leftarrow \bot$ is added.

No Derivation Through Double Negation (doubleNeg). A positive conclusion can be derived from double negation using two conditionals under the WCS. It appears to be the case that humans do not reason in such a way (see [7]). Hence, we block them with the help of abnormalities.

4   Quantified Statements as Logic Programs

Based on the principles and encoding aspects presented in Sects. 2 and 3.6, we encode the quantified statements into logic programs. The programs are specified using the predicates y and z and depend on the figures shown in Table 2, where yz can be replaced by ab, ba, cb, or bc. Here, all principles regarding a premise are applied. However, we will later assume different clusters of reasoners, some of which do not apply certain principles (see Sect. 5). The clauses associated with principles that are not applied are removed for such clusters.

4.1   All y Are z (Ayz)

All y are z is represented by $\mathcal{P}_{Ayz}$, which consists of the following clauses:

$$
\begin{array}{ll}
z(X) \leftarrow y(X) \wedge \neg ab_{yz}(X). & \text{(conditionals \& licenses)} \\
ab_{yz}(X) \leftarrow \bot. & \text{(licenses)} \\
y(o) \leftarrow \top. & \text{(import)} \\
ab_{yz}(X) \leftarrow \mathsf{ctxt}(z'(X)). & \text{(contraposition \& licenses \& deliberateGen)} \\
y'(X) \leftarrow \neg z(X) \wedge \neg ab_{zy}(X). & \text{(contraposition \& conditionals \& licenses)} \\
ab_{zy}(X) \leftarrow \bot. & \text{(contraposition \& licenses)} \\
y(X) \leftarrow \neg y'(X) \wedge \neg ab_{nyy}(X). & \text{(contraposition \& transformation \& licenses)}
\end{array}
$$

As contraposition has been applied, we have to add the integrity constraint $\mathsf{U} \leftarrow y(X) \wedge y'(X)$. We obtain $\mathcal{M}_{\mathcal{P}_{Ayz}} = \langle \{y(o), z(o)\}, \{ab_{yz}(o)\} \rangle$. Remember that we want to construct pairs of syllogistic premises. Sometimes, if a premise of A mood is combined with a premise of E or O mood (see Sects. 4.2 and 4.4), then $z'(X)$ appearing in the body of the fourth clause becomes the negation of $z(X)$. Otherwise, any ground instance of $z'(X)$ is unknown and, consequently, $\text{ctxt}(z'(X))$ is false in this case. The necessity of the fourth clause and the usage of the ctxt operator is discussed in the example presented in Sect. 5.2.

4.2   No y Is z (Eyz)

No y is z is represented by $\mathcal{P}_{Eyz}$, which consists of the following clauses:

$$z'(X) \leftarrow y(X) \wedge \neg ab_{ynz}(X). \qquad \text{(transformation \& licenses)}$$
$$ab_{ynz}(X) \leftarrow \bot. \qquad \text{(licenses)}$$
$$z(X) \leftarrow \neg z'(X) \wedge \neg ab_{nzz}(X). \qquad \text{(transformation \& licenses)}$$
$$y(o_1) \leftarrow \top. \qquad \text{(import)}$$
$$ab_{nzz}(o_1) \leftarrow \bot. \qquad \text{(licenses \& doubleNeg)}$$
$$y'(X) \leftarrow z(X) \wedge \neg ab_{zny}(X). \qquad \text{(converse \& transformation \& licenses)}$$
$$ab_{zny}(X) \leftarrow \bot. \qquad \text{(converse \& licenses)}$$
$$y(X) \leftarrow \neg y'(X) \wedge \neg ab_{nyy}(X). \qquad \text{(converse \& transformation \& licenses)}$$
$$z(o_2) \leftarrow \top. \qquad \text{(converse \& import)}$$
$$ab_{nyy}(o_2) \leftarrow \bot. \qquad \text{(converse \& licenses \& doubleNeg)}$$

The integrity constraints $\mathsf{U} \leftarrow z(X) \wedge z'(X)$ and $\mathsf{U} \leftarrow y(X) \wedge y'(X)$ must be added. Iterating $\Phi_{\mathcal{P}_{Eyz}}$ we obtain $\mathcal{M}_{\mathcal{P}_{Eyz}} = \langle \{y(o_1), z'(o_1), z(o_2), y'(o_2)\}, \{ab_{ynz}(o_1), ab_{nzz}(o_1), z(o_1), ab_{zny}(o_2), ab_{nyy}(o_2), y(o_2)\}\rangle$.

4.3   Some y Are z (Iyz)

Some y are z is represented by $\mathcal{P}_{Iyz}$, which consists of the following clauses:

$$z(X) \leftarrow y(X) \wedge \neg ab_{yz}(X). \qquad \text{(conditionals \& licenses)}$$
$$ab_{yz}(o_1) \leftarrow \bot. \qquad \text{(unknownGen \& licenses)}$$
$$y(o_1) \leftarrow \top. \qquad \text{(import)}$$
$$y(o_2) \leftarrow \top. \qquad \text{(unknownGen)}$$
$$ab_{yz}(X) \leftarrow \text{ctxt}(z'(X)). \qquad \text{(licenses \& deliberateGen)}$$
$$ab_{yz}(o_2) \leftarrow \mathsf{U}. \qquad \text{(licenses \& deliberateGen)}$$
$$y(X) \leftarrow z(X) \wedge \neg ab_{zy}(X). \qquad \text{(converse \& conditionals \& licenses)}$$
$$ab_{zy}(o_3) \leftarrow \bot. \qquad \text{(converse \& licenses \& unknownGen)}$$
$$z(o_3) \leftarrow \top. \qquad \text{(converse \& import)}$$
$$z(o_4) \leftarrow \top. \qquad \text{(converse \& unknownGen)}$$
$$ab_{zy}(X) \leftarrow \text{ctxt}(y'(X)). \qquad \text{(converse \& licenses \& deliberateGen)}$$
$$ab_{zy}(o_4) \leftarrow \mathsf{U}. \qquad \text{(converse \& licenses \& deliberateGen)}$$

We obtain $\mathcal{M}_{\mathcal{P}_{Iyz}} = \langle \{y(o_1), y(o_2), z(o_1)\}, \{ab_{yz}(o_1)\}\rangle$. One should observe that $ab_{yz}(o_2)$ is an unknown assumption in $\mathcal{P}_{Iyz}$ and, hence, $\mathcal{M}_{\mathcal{P}_{Iyz}}(z(o_2)) = \mathsf{U}$.

4.4   Some y Are Not z (Oyz)

Some y are not z is represented by $\mathcal{P}_{Oyz}$ which consists of the following clauses:

$z'(X) \leftarrow y(X) \wedge \neg ab_{ynz}(X).$	(conditionals & transformation & licenses)
$ab_{ynz}(o_1) \leftarrow \bot.$	(unknownGen & licenses)
$z(X) \leftarrow \neg z'(X) \wedge \neg ab_{nzz}(X).$	(transformation & licenses)
$y(o_1) \leftarrow \top.$	(import)
$y(o_2) \leftarrow \top.$	(unknownGen)
$ab_{nzz}(o_1) \leftarrow \bot.$	(doubleNeg & licenses)
$ab_{nzz}(o_2) \leftarrow \bot.$	(doubleNeg & licenses)

We have to add the integrity constraint $U \leftarrow z(X) \wedge z'(X)$ and obtain $\mathcal{M}_{\mathcal{P}_{Oyz}} = \langle \{y(o_1), y(o_2), z'(o_1)\}, \{ab_{ynz}(o_1), ab_{nzz}(o_1), ab_{nzz}(o_2), z(o_1)\}\rangle$.

4.5   Entailment of Conclusions from Pairs of Syllogistic Premises

Based on the applied principles of the previous section, we specify when $\mathcal{M}_{\mathcal{P}}$ entails a conclusion, where yz is to be replaced by ac or ca.

Ayz (all) $\mathcal{P} \models Ayz$ iff there exists an object o such that $\mathcal{P} \models_{wcs} y(o)$ and for all objects o we find that if $\mathcal{P} \models_{wcs} y(o)$ then $\mathcal{P} \models_{wcs} z(o)$.

Eyz (no) $\mathcal{P} \models Eyz$ iff there exists an object o_1 such that $\mathcal{P} \models_{wcs} y(o_1)$ and for all objects o_1 we find that if $\mathcal{P} \models_{wcs} y(o_1)$ then $\mathcal{P} \models_{wcs} \neg z(o_1)$ and there exists an object o_2 such that $\mathcal{P} \models_{wcs} z(o_2)$ and for all objects o_2 we find that if $\mathcal{P} \models_{wcs} z(o_2)$ then $\mathcal{P} \models_{wcs} \neg y(o_2)$.

Iyz (some) $\mathcal{P} \models Iyz$ iff there exists an object o_1 such that $\mathcal{P} \models_{wcs} y(o_1) \wedge z(o_1)$ and there exists an object o_2 such that $\mathcal{P} \models_{wcs} y(o_2)$ and $\mathcal{P} \not\models_{wcs} z(o_2)$ and there exists an object o_3 such that $\mathcal{P} \models_{wcs} z(o_3) \wedge y(o_3)$ and there exists an object o_4 such that $\mathcal{P} \models_{wcs} z(o_4)$ and $\mathcal{P} \not\models_{wcs} y(o_4)$.

Oyz (Some Are Not) $\mathcal{P} \models Oyz$ iff there exists an object o_1 such that $\mathcal{P} \models_{wcs} y(o_1) \wedge \neg z(o_1)$ and there exists an object o_2 such that $\mathcal{P} \models_{wcs} y(o_2)$ and $\mathcal{P} \not\models_{wcs} \neg z(o_2)$.

NVC When no previous conclusion can be derived, no valid conclusion holds.

4.6   Accuracy of Predictions

We have nine different answer possibilities for each of the 64 pairs of syllogistic premises: Aac, Eac, Iac, Oac, Aca, Eca, Ica, Oca and NVC. For every pair of syllogistic premises, we define two lists of length nine for the predictions of the WCS and for the participants' answers, where the first element represents Aac, the second element represents Eac, and so forth. When Aac is predicted under the WCS (or the majority's conclusions) for a given pair of syllogistic premises, then the value of the first element of this list is a 1, otherwise it is a 0, and the same holds for the other eight elements in the list. Given

$$\text{COMP}(i) = \begin{cases} 1 & \text{if both lists have the same value for the } i\text{th element} \\ 0 & \text{otherwise} \end{cases}$$

the matching percentage of this pair of syllogistic premises is then computed by $\sum_{i=1}^{9} \text{COMP}(i)/9$.

5 Clusters and Heuristics

We understand clusters of human reasoners in terms of principles or heuristics. Each cluster is a group of humans that applies the same principles or heuristics. When identifying such clusters, e.g., among the participants in [7], the principles or heuristics used by a single cluster should lead to a significant answer for the pair of syllogistic premises in question. As the answers of all participants have been accumulated in the meta-analysis, the combined answers of all clusters should exactly correspond to the significant answers for that pair of syllogistic premises.

5.1 Basic Principles

Basic principles are assumed to be applied by all reasoners, regardless of any cluster. These are conditionals, licenses, import, and unknownGen. Note that they are not necessarily applicable to every pair of syllogistic premises: unknownGen may only be used for premises with an existential mood.

5.2 Advanced Principles and Clusters

Advanced principles are assumed by some but not all humans, making them the starting point for clusters. Advanced principles considered in this paper are converse, deliberateGen, contraposition, and searchAlt, but there may exist more. When two individuals differ in the sense that one applies such a principle and the other one does not, we assume that they belong to different clusters.

As an example, consider AO3 introduced in Sect. 1. According to the encoding described in Sect. 4, $\mathcal{P}_{\text{AO3,basic}}$ represents the logic program for AO3, where only the basic principles are applied:

$$
\begin{array}{lll}
b(X) \;\;\leftarrow a(X) \wedge \neg ab_{ab}(X). & b'(X) \leftarrow c(X) \wedge \neg ab_{cnb}(X). & c(o_3) \;\;\;\leftarrow \top. \\
ab_{ab}(X) \leftarrow \bot. & c(o_2) \leftarrow \top. & ab_{nbb}(o_2) \leftarrow \bot. \\
a(o_1) \;\;\;\leftarrow \top. & b(X) \;\;\leftarrow \neg b'(X) \wedge \neg ab_{nbb}(X). & ab_{cnb}(o_2) \leftarrow \bot. \\
& & ab_{nbb}(o_3) \leftarrow \bot.
\end{array}
$$

We obtain

$$
\mathcal{M}_{\mathcal{P}_{\text{AO3,basic}}} = \langle \; \{ \; \boxed{a(o_1)} \,, b(o_1), \; \boxed{c(o_2)} \,, \; \boxed{c(o_3)} \,, b'(o_2) \},
$$
$$
\{ ab_{ab}(o_1), ab_{ab}(o_2), ab_{ab}(o_3), ab_{cnb}(o_2), ab_{nbb}(o_2), ab_{nbb}(o_3) \} \rangle.
$$

The highlighted atoms are relevant for conclusions: NVC follows. Note that $ab_{ab}(o_i)$ is false for all o_i, $1 \leq i \leq 3$. If additionally contraposition is used,

$$
\begin{aligned}
\mathcal{P}_{\text{AO3,contra}} = \mathcal{P}_{\text{AO3,basic}} \; \cup \; \{ a'(X) \leftarrow \neg b(X) \wedge \neg ab_{ba}(X), \; ab_{ba}(X) \leftarrow \bot, \\
a(X) \leftarrow \neg a'(X) \wedge \neg ab_{naa}(X), \; ab_{ab}(X) \leftarrow \text{ctxt}(b'(X)) \},
\end{aligned}
$$

is considered that has another clause where $ab_{ab}(X)$ is in the head. We obtain

$$
\mathcal{M}_{\mathcal{P}_{\text{AO3,contra}}} = \langle \; \{ \; \boxed{a(o_1)} \,, ab_{ab}(o_2), b(o_1), \; \boxed{c(o_2)} \,, \; \boxed{c(o_3)} \,, a'(o_2), b'(o_2) \},
$$
$$
\{ \; \boxed{a(o_2)} \,, ab_{ab}(o_1), ab_{ab}(o_3), ab_{cnb}(o_2), ab_{nba}(o_1), ab_{nba}(o_2),
$$
$$
ab_{nba}(o_3), ab_{nbb}(o_2), ab_{nbb}(o_3), b(o_2), a'(o_1) \} \rangle.
$$

Again, the relevant atoms are highlighted. $\mathcal{M}_{\mathcal{P}_{AO3,\text{contra}}}$ entails the conclusion Oca: As $c(o_2)$ is true, $b'(o_2)$ is true, therefore $\text{ctxt}(b'(o_2))$ is true. This in turn makes $ab_{ab}(o_2)$ true, and accordingly $b(o_2)$ has to be false. But then, $a'(o_2)$ can be derived true, which finally makes $a(o_2)$ false.

One should observe that $b'(o_1)$ is unknown in $\mathcal{M}_{\mathcal{P}_{AO3,\text{contra}}}$. Hence, $\text{ctxt}(b'(o_1))$ is false and, consequently, $ab_{ab}(o_1)$ is false as well. Together with $a(o_1)$ being true we obtain that $b(o_1)$ is true. The latter is needed to correctly implement the first premise, *all a are b*, in AO3. Without the ctxt operator, $b'(o_1)$ would be unknown and, consequently $ab_{ab}(o_1)$ as well as $b(o_1)$ would be unknown as well violating the premise *all a are b*.

Assuming two clusters of people whose reasoning process differs in the application of the contraposition principle, we unite the conclusions predicted for the clusters and obtain {Oca, NVC}. These are exactly the significant answers reported in [7].

In order to represent the principles leading to a conclusion, multinomial processing trees (MPTs) [11] are used. They have been suggested for modeling cognitive theories because they represent cognitive processes as probabilistic procedures, thus being able to predict multiple answers and even their quantitative distribution [10]. We set the latent states (inner nodes) of the MPTs to the decisions whether to use certain principles and put the corresponding conclusions in the leaves. The MPT for AO3 based on the clustering described above is presented in Fig. 1. The parameter $p_{\text{contraposition}}$ models the probability that an individual applies the contraposition principle and, therefore, belongs to the corresponding cluster. It can be trained from experimental data with algorithms like expectation-maximization [6]. Note that the MPT in Fig. 1 is not complete in the sense that it cannot predict all possible conclusions for AO3. This issue is addressed below.

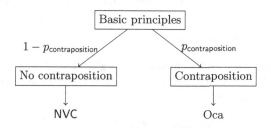

Fig. 1. The MPT for AO3.

5.3 Heuristic Strategies

Some theories suggest that some humans do not reason at all to solve syllogistic reasoning tasks, but rely on heuristics such as the atmosphere bias [14] or the matching bias [13]. Such heuristics are simple rules that state what conclusions are likely depending on certain features of the premises, e.g., mood or figure.

Some of the participants' answers presented in [7], that are given by a small amount of people (less then 5%), but also some significant ones, are not (yet) explainable by the WCS. A plausible explanation for that is that these people simply guess or use one of the heuristics mentioned below (educated guess).

A *generative approach* to model this behavior can be based on MPTs. The MPT for a random guess can lead to all nine conclusions. MPTs for a particular heuristic strategy only take into account the valid conclusions under the corresponding theory. For the atmosphere bias, universal and affirmative conclusions are excluded when one of the premises is existential or negative, respectively. In the case of identical moods, the conclusion must have this mood as well. For the matching bias, the following order from the most to the least conservative quantifier is defined on moods:

$$E > O = I > A.$$

A conclusion may not be answered if it is *less conservative* than one of the premises with respect to that order. We have also observed biased conclusions in the data of [7] that may be explained by the following heuristic strategy: F or almost all pairs of syllogistic premises with Fig. 1, Xac is answered, while the answer Xca is not given at all, where X is the least conservative mood from the premises that is still allowed under the matching strategy (O is preferred over I).

As an alternative to generating the answers given by a cluster of guessers using MPTs, the following inverse process can be considered: predictions of the WCS that are not in accordance with a particular heuristic strategy are not given by a cluster using that strategy. In the *filtering approach*, these conclusions are suppressed in the predictions. If no conclusion remains, NVC is answered instead. As it is likely that some participants do not use logic [13], such clusters must be modeled under the WCS by using the generative or the filtering approach. As a consequence, MPTs can construct a prediction for all answer possibilities.

5.4   A Clustering Approach

Based on the principles and heuristic strategies described above, the participants of [7] have been partitioned into three reasoning clusters and two clusters applying heuristic strategies:

1. Basic principles, searchAlt, and converse for I.
2. Basic principles, converse for I and deliberateGen.
3. Basic principles, converse for I, E, and contraposition for A.
4. Matching strategy.
5. Biased conclusions in figure 1.

Abduction was only used in one cluster because of the computational effort it requires. Although it would be interesting to model this principle for different clusters, the impact would be very small. This is because converse is the only advanced principle that adds existential imports, which we currently consider as atoms for observations. According to the results of [8], abduction has the same results independent of whether only the converse I mood or both the converse I and E mood are used. The matching strategy was implemented using the filtering approach. The *biased conclusions in figure 1* heuristics was implemented using the generative approach such that its prediction overwrites the answers of other clusters, except NVC.

Table 4. Comparison of the WCS with other cognitive theories. The participants' answers are highlighted.

	Participants	PSYCOP	Verbal Models	Mental Models	Conversion	WCS
AO3	Oca	Oca	Oca	Oca	Oca	Oca
	NVC	Ica Iac	NVC	NVC Oac	NVC	NVC
Overall	100 %	77 %	84 %	78 %	83 %	92 %

5.5   Evaluation

We evaluate the predictions of the WCS based on the clustering approach described in Sect. 5.4. For that, we combine the answers of all clusters and compared them with both the data of humans and the predictions of other cognitive theories presented in [7]. In that study, the results of six psychological experiments on syllogistic reasoning were aggregated and compared with twelve well-known cognitive theories. In Table 4, it can be seen that the WCS predicts the same answers for AO3 as the majority of humans, but some other theories fail to do so. For the overall evaluation, the accuracy is computed as described in Sect. 4.6. Here the WCS clearly stands out against the other theories, but to be fair, we must also admit that we compare a relatively new theory to the best theories of 2012. The WCS predicts the participants' answers in [7] correctly for 32 out of the 64 pairs of syllogistic premises. For 20 cases there is one incorrect prediction, for 11 cases there are two and for one case there are three mismatches. The overall match between the predictions of the WCS and the answers of the participants is 92%.

6   Implementation

The goal of our implementation is to automate the process of evaluating a certain clustering. This is crucial, because as stated above, the number of possible clusterings grows exponentially with the number of principles. We want to be able to evaluate new candidates for an optimal clustering as fast as possible.

We have developed a modular, declarative implementation, which consists of two parts: An implementation of the $\Phi_{\mathcal{P}}$ operator to compute the least fixed point of a given program $\mathcal{P}$, and a framework that generates logic programs from an abstract representation of principles and evaluate the entailed conclusions.

6.1   Computing the Least Fixed Point of $\Phi_\mathcal{P}$

The least fixed point of $\Phi_\mathcal{P}$ is computed in Prolog. The implementation receives a program $\mathcal{P}$ – written in Prolog – as input and processes it in two phases. The output is an interpretation $\langle I^\top, I^\perp \rangle$ of wc $\mathcal{P}$ represented as two lists corresponding to $I^\top$ and $I^\perp$. The input program $\mathcal{P}$ is first grounded to obtain g$\mathcal{P}$ and, secondly, computes the least fixed point of $\Phi_\mathcal{P}$ starting with the empty interpretation $\langle \emptyset, \emptyset \rangle$. Recall that $\Phi_\mathcal{P}$ operates directly on g$\mathcal{P}$. The context operator is implemented such that contextual logic programs can be handled. However, there is a problem: if a contextual logic program $\mathcal{P}$ contains a cycle, then the least fixed point of $\Phi_\mathcal{P}$ may not exist. Consider the following quantified statements:

$$\textit{All a are b.} \qquad \textit{No b is c.} \qquad \text{(AE1)}$$

Assume that additionally to the basic principles we apply for each quantified statement the advanced principles converse, deliberateGen, and contraposition. The corresponding program consists of the following clauses:

$$b(X) \leftarrow a(X) \wedge \neg \; ab_{ab}(X).$$
$$ab_{ab}(X) \leftarrow \perp.$$
$$a(o_1) \leftarrow \top.$$
$$ab_{ab}(X) \leftarrow \mathsf{ctxt}(\; b'(X)\;).$$
$$a'(X) \leftarrow \neg b(X) \wedge \neg ab_{ba}(X).$$
$$ab_{ba}(X) \leftarrow \perp.$$
$$a(X) \leftarrow \neg a'(X) \wedge \neg ab_{naa}(X).$$

$$c'(X) \leftarrow b(X) \wedge \neg ab_{nbc}(X).$$
$$b(o_2) \leftarrow \top.$$
$$ab_{nbc}(X) \leftarrow \perp.$$
$$c(X) \leftarrow \neg \; c'(X) \wedge \neg ab_{ncc}(X).$$
$$ab_{ncc}(o_2) \leftarrow \perp.$$
$$b'(X) \leftarrow \neg \; c(X) \wedge \neg ab_{ncb}(X).$$
$$ab_{ncb}(X) \leftarrow \perp.$$
$$b(X) \leftarrow \neg b'(X) \wedge \neg ab_{nbb}(X).$$

Consider the highlighted atoms: Note the cycle $b' > c > c' > b > ab_{ab} > b'$ where $A > B$ if A is an atom in the head of a rule and B is an atom that occurs in the body of that rule. As b' is an argument of the context operator and is part of the cycle, this program does not admit a least fixed point. When modeling clusters, we must ensure that the logic program resulting from the applied principles do not contain such cycles. This is guaranteed for the clusters given in Sect. 5.4.

6.2   Computing the Predictions for a Cluster of Reasoners

The evaluation of a cluster is written in Haskell. A run consists of four phases:

1. Generate program $\mathcal{P}$ of the pair of syllogistic premises using the principles.
2. Call the Prolog implementation to compute the least fixed point of $\Phi_\mathcal{P}$.
3. Extract the conclusions entailed by the least fixed point of $\Phi_\mathcal{P}$.
4. Compare the conclusions with the participants' answers and output score.

The Haskell program contains definitions of datatypes for all entities occurring in the programs, i.e., truth values, atoms, literals, and clauses. These entities are built recursively on each other and have functions for conversion into Prolog. Principles are implemented as functions that return their corresponding clause representation. The source code of the unknownGen principle is as follows:

```
unknownGen = Principle {
    apply = \m f -> m == MI || m == MO,
    clauseRep = \ y z prf -> [clause (atom y) [top] (prf ++ "ug")]
}
```

where the first line states that the principle is applied to negative moods (I and O) and the second line states that the clause has the form $y(prf\mathrm{ug}) \leftarrow \top$, where prf is an identifier for objects of the clause. Using this abstraction, clusters are written as lists of 'principle functions' and are thus valid Haskell source code by themselves. As an example, consider the definition of the basic cluster which uses the basic principles and the converse principle for mood I:

```
basicCluster = Cluster {
    principles = basicPrinciples ++ map converseI basicPrinciples,
    ...
}
```

Here, basicPrinciples is defined as list of principles (those we called *basic* in Sect. 5.1). Of course, one consequence is that the user of our implementation has to be familiar with Haskell. However, there are two main advantages of using Haskell source code as a representation. Firstly, many principles are part of a certain subset of the pair of syllogistic premises (e.g., the unknown generalization principle is used for all premises with an existential mood). These connections can be modeled precisely and without redundancy in source code. This can be seen in the example above, where converseI is implemented as a function that takes a principle as argument and returns the corresponding principle for the converse premise. Secondly, because Haskell is a compiled language, the representation of the pair of syllogistic premises itself is compiled. Therefore, a representation is automatically checked and the program does not crash due to an error, which would not be the case if e.g., a string representation was used.

The Prolog representation of the program results from a function converting sets of clauses to a string and is written into a file. Then, the Prolog implementation of the previous program is called to compute the least fixed point of $\Phi_{\mathcal{P}}$, which is again written to a file. After completion that file is parsed and the conclusions are extracted with respect to the definitions given in Sect. 4.5. Our heuristic filters—implemented as post-processing functions—are applied to these conclusions. This process is done for all 64 pairs where the conclusions are compared with the participants' answers and the score of the cluster is computed.

Until now, we have only described the evaluation of a single cluster, although a clustering consists of the combined answers of all clusters. For this purpose, a list of clusters is specified, where the program computes the predictions for each cluster, combines them, and compares the results with the participants' answers.

7   Conclusions

We have successfully extended the approach in [1,8] by introducing two new principles and by applying a clustering approach to model individual differences in human reasoning. This takes into account that some people may not reason at all, but guess or apply heuristic strategies. The clustering presented in Sect. 5.4 is currently the best one but possibly not the optimal one. However, due to the combinatorial explosion,[2] it is difficult to find the global optimum. Furthermore, programs based on certain principles considered for some moods, might not have a least fixed point, as they contain cycles with respect to the ctxt operator. This must be taken into account when selecting the principles for a clustering. Finally, we have applied multinomial processing trees to model that different principles lead to different conclusions. This information is lost if the data containing the predictions for all clusters is aggregated. If we would have more insight about the patterns participants opted for, we could model single pair of syllogistic premises by multinomial process trees instead of fitting them to the overall results.

Future work might allow us to identify and understand why humans within a cluster come to certain conclusions. Accordingly, if it is known which principles they apply, it should be possible to predict their answers.

References

1. Costa, A., Dietz Saldanha, E.-A., Hölldobler, S.: Monadic reasoning using weak completion semantics. In: Hölldobler, S., Malikov, A., Wernhard, C. (eds.) Proceedings of the Young Scientist's Second International Workshop on Trends in Information Processing (YSIP2) 2017, pp. 45–54. CEUR Workshop Proceedings (2017)
2. Dietz Saldanha, E.-A., Hölldobler, S., Pereira, L.M.: Contextual reasoning: usually birds can abductively fly. In: Balduccini, M., Janhunen, T. (eds.) LPNMR 2017. LNCS (LNAI), vol. 10377, pp. 64–77. Springer, Cham (2017). https://doi.org/10.1007/978-3-319-61660-5_8
3. Grice, H.P.: Logic and conversation. In: Cole, P., Morgan, J.L. (eds.) Syntax and Semantics, vol. 3. Academic Press (1975)
4. Hölldobler, S.: Weak completion semantics and its applications in human reasoning. In: Furbach, U., Schon, C. (eds.) Proceedings of the Workshop on Bridging the Gap between Human and Automated Reasoning on the 25th International Conference on Automated Deduction, CEUR Workshop Proceedings, vol. 1412, pp. 2–16. CEUR-WS.org. (2015)
5. Hölldobler, S., Kencana Ramli, C.D.P.: Logic programs under three-valued Łukasiewicz semantics. In: Hill, P.M., Warren, D.S. (eds.) ICLP 2009. LNCS, vol. 5649, pp. 464–478. Springer, Heidelberg (2009). https://doi.org/10.1007/978-3-642-02846-5_37
6. Hu, X., Batchelder, W.H.: The statistical analysis of general processing tree models with the em algorithm. Psychometrika **59**(1), 21–47 (1994)

---

[2] For n principles, there are up to 2^n possible clusters. Additionally, it is unknown if the current set of principles is already complete.

7. Khemlani, S., Johnson-Laird, P.N.: Theories of the syllogism: a meta-analysis. Psychol. Bull. **138**, 427–457 (2012)

8. Costa, A., Dietz Saldanha, E.-A., Hölldobler, S., Ragni, M.: A computational logic approach to human syllogistic reasoning. In: Gunzelmann, G., Howes, A., Tenbrink, T., Davelaar, E.J. (eds.) Proceedings of the 39th Annual Conference of the Cognitive Science Society, Austin, TX, 2017, pp. 883–888. Cognitive Science Society (2017)

9. Ragni, M., Dietz, E.-A., Kola, I., Hölldobler, S.: Two-valued logic is not sufficient to model human reasoning, but three-valued logic is: a formal analysis. In: Schon, C., Furbach, U. (eds.) Proceedings of the Workshop on Bridging the Gap between Human and Automated Reasoning co-located with 25th International Joint Conference on Artificial Intelligence IJCAI, CEUR Workshop Proceedings, vol. 1651, pp. 61–73. CEUR-WS.org. (2016)

10. Ragni, M., Singmann, H., Steinlein, E.-M.: Theory comparison for generalized quantifiers. In: CogSci (2014)

11. Riefer, D.M., Batchelder, W.H.: Multinomial modeling and the measurement of cognitive processes. Psychol. Rev. **95**(3), 318–339 (1988)

12. Stenning, K., van Lambalgen, M.: Human Reasoning and Cognitive Science. A Bradford Book. MIT Press, Cambridge (2008)

13. Wetherick, N.E., Gilhooly, K.J.: 'Atmosphere', matching, and logic in syllogistic reasoning. Curr. Psychol. **14**(3), 169–178 (1995)

14. Woodworth, R.S., Sells, S.B.: An atmosphere effect in formal syllogistic reasoning. J. Exper. Psychol. **18**(4), 451 (1935)

Functional and Logic Programming

Concolic Testing of Functional Logic Programs

Jan Rasmus Tikovsky[✉]

Institut für Informatik, CAU Kiel, 24098 Kiel, Germany
jrt@informatik.uni-kiel.de

Abstract. In the last years, concolic testing, a technique combining concrete and symbolic execution for the automated generation of test cases, has gained increasing popularity. Concolic testing tools are initialized with expressions on concrete input data. But instead of just evaluating them, they additionally collect symbolic information along specific execution paths. This information can be used to systematically compute alternative inputs exploring yet unvisited paths. In this way, test cases can be generated covering all branches of a given program. The first concolic testing tools have been developed for imperative languages analyzing code at a very low level. Recently, there have been also some approaches investigating the concolic execution of declarative languages. In this work, we discuss the application of concolic testing to the functional logic language Curry. More precisely, we present *ccti*, a concolic interpreter which is adapted for the automated generation of test cases for both purely functional and non-deterministic programs.

1 Introduction

There are several methods to verify the correctness of programs. Among these formal program verification has the most significant relevance. But as proving the correctness of programs is a rather difficult and time consuming task, testing has become the most established approach to ensure the reliability of software. In fact, program testing itself became a wide area of research over the last decade resulting in various approaches.

In general, we distinguish between testing in the large and testing in the small. The former includes the testing of complete systems as well as the verification of interfaces between larger components, while the latter is directed to minor parts of programs like one module or even only a single function.

Furthermore, regarding the consideration of source code, software testing can be divided into two categories, namely black-box and glass-box testing. As the name implies, tests of the former category treat the software to be tested like a black box ignoring its concrete implementation completely and deducing test cases from specifications. Random testing and property-based testing, falls within this category. Property-based testing uses random input data to produce results which are then matched with previously specified properties.

© Springer Nature Switzerland AG 2018
D. Seipel et al. (Eds.): DECLARE 2017, LNAI 10997, pp. 169–186, 2018.
https://doi.org/10.1007/978-3-030-00801-7_11

QuickCheck [5] for Haskell or QuviQ QuickCheck[1] for Erlang are examples of property-based testing libraries for functional languages. But also purely logic languages like Prolog and functional logic languages like Curry provide tools for property testing, namely PrologCheck [2] and CurryCheck [11].

Glass-box testing, on the other hand, works on the source code level. By the selection of input data, execution paths are followed through the code to determine appropriate outputs. Often this process is repeated until certain code coverage criteria are met. Since glass-box testing is a systematic approach, it is well-suited for automation. Examples of glass-box testing include test case generation based on symbolic execution. In this process, a program is interpreted using symbolic values for inputs instead of concrete data producing constraints on those symbols for all conditional branches in the program. Applying constraint solvers, these so called path constraints can be solved to compute actual input data driving execution along the associated path.

In the last years, a combination of concrete and symbolic execution, called concolic execution, has gained more and more popularity. We explain the basic idea behind concolic testing with the following example program.

```
nthElem []          _         = Nothing
nthElem (x : xs) n | n == 0 = Just x
                   | n >  0 = nthElem xs (n - 1)
```

List. 1. Selection of the n-th element of a list

The listing shows the definition of a Curry function to select the n-th element of a polymorphic list. This definition distinguishes three different cases via pattern matching and guards: If the given list is empty, Nothing is returned. Considering a non-empty list either the first list element is returned or the function is called recursively depending on whether the index is 0 or a positive integer number.

The objective of automated testing tools is to find enough test cases to fully cover every branch of a function at least once. For the given example two test cases would be sufficient, i.e., one using an empty list and one using a list with at least two elements and an index greater than 0 and smaller than the length of the list.

For this purpose, concolic testing tools start with some concrete inputs. While evaluating a function call with these inputs, concolic testing tools additionally collect symbolic information describing the branch decisions which are made along that execution path. These decisions are also denoted as path constraints, since input data has to satisfy them to drive execution along that path.

Concolic testing tools aim at negating such path constraints systematically and solving them to produce inputs which drive evaluation along alternative execution paths. Repeating this process automatically generates test cases covering all program branches.

For instance, during the concrete execution of "nthElem [42] 0", we additionally consider the symbolic expression nthElem xs n with xs and n being symbolic variables. The concrete expression can be evaluated to Just 42 by applying

---

[1] http://www.quviq.com/products/erlang-quickcheck/.

the second rule of nthElem. During pattern matching, a branch decision is made constraining the symbolic variable xs to a non-empty list. Furthermore, the evaluation of the first guard of the second rule constrains the symbolic variable n to be equal to 0. By negating these constraints, we receive constraints associated with alternative execution paths. For example, the negation of the first path condition constrains xs to be the empty list, thus, driving execution along a different execution path, namely the one represented by the first rule of nthElem. This process is repeated until all paths of the associated symbolic execution tree have been visited and, thus, all branches of nthElem are covered.

The first concolic testing tools were developed for imperative languages. Examples include DART [10] and CUTE [18] for C, and jCUTE [17] for Java. Recently, concolic execution has found its way into declarative programming languages. For the functional language Erlang there are two tools which apply a program instrumentation to collect symbolic information, namely [9] and [16]. Moreover, in [14] and [15] a method for concolic testing in Prolog is presented. Regarding the functional logic language Curry, Fischer and Kuchen [8] discuss an approach which uses narrowing to generate test cases from uninstantiated function arguments systematically.

In this work, we propose *ccti* (Curry Concolic Testing Interpreter)[2], a tool for automated concolic execution of Curry programs. To the best of our knowledge, concolic testing so far has not been applied to functional logic programs. We present an augmented semantics for Curry's simplified core language FlatCurry which enables the additional collection of symbolic information during concrete evaluation. This symbolic information is used to generate path constraints. By negating these constraints systematically and applying an SMT solver, namely Z3 [6], we produce input data directing the execution to yet unexplored program paths. Furthermore, we present a simple search strategy for the selection of the path constraint to be negated next. Although *ccti*'s search interface supports the implementation of different coverage criteria, in this work we focus on branch coverage.

Our work is based on approaches applying concolic testing to purely functional languages and demonstrates that some of the ideas proposed in these approaches can be applied to functional logic languages as well. For instance, we also use a simplified core language which facilitates the identification of program branches, and thus the collection of path constraints. Moreover, *ccti* provides a search strategy to explore alternative, yet unvisited execution paths which is very similar to the one presented in [9]. In contrast to the concolic testing tools for purely functional languages mentioned above, we use an interpretation-rather than an instrumentation-based approach. This is due to the fact that the combination of non-deterministic computations and sharing of common subexpressions in Curry complicates the implementation of a semantics-preserving code instrumentation. Contrary to the narrowing-based approach presented in [8], *ccti* enables the generation of test cases for programs including primitive types like integers or floats. While narrowing on those primitive types can only

---

[2] https://www-ps.informatik.uni-kiel.de/~jrt/forschung/ccti.html.

be applied by using alternative, data constructor-based representations of integers and floats, we can simply reuse their original representation by applying suitable theories of the SMT solver.

The rest of this paper is structured as follows: In Sect. 2 we describe the functional logic language Curry and its simplified core language FlatCurry. Section 3 gives a brief introduction to satisfiability modulo theories (SMT) and the SMT-LIB library. The general idea of concolic testing of FlatCurry programs is explained in Sect. 4. Afterwards, we present a variant of the natural semantics of FlatCurry programs augmented for concolic testing. In Sect. 5 we take a closer look at the search algorithm applied to investigate the symbolic information which has been collected during concolic execution. Section 6 presents parts of the implementation of *ccti*. Finally, we discuss the applicability of *ccti* considering some practical examples in Sect. 7 before we conclude in Sect. 8.

2 Curry

Curry is a declarative programming language integrating well-known features from functional programming, like higher-order functions and lazy evaluation, as well as elements of logic programming, like non-determinism and computations with partial information. We will give only a short overview here. For a detailed introduction we refer to [13].

Curry's syntax is very similar to that of the functional programming language Haskell. Curry supports the declaration of algebraic data types via the keyword `data`. Identifiers of types and data constructors start with an uppercase letter, whereas variable and function names usually begin with a lowercase letter.

For instance, the `Maybe` type representing optional values in Curry which we used in List. 1 is defined as follows.

```
data Maybe a = Nothing | Just a
```

Functions are defined via rules and pattern matching. In contrast to Haskell, Curry supports the definition of non-deterministic operations by specifying overlapping rules. The following listing shows the definition of an operation which inserts an element in a list at an arbitrary position.

```
insertND x []     = [x]
insertND x (y:ys) = x : y : ys
insertND x (y:ys) = y : insertND x ys
```

Consider the expression "`insertND 42 [1,2]`". Evaluating this expression in a Curry system will yield three non-deterministic results, namely `[42,1,2]`, `[1,42,2]` and `[1,2,42]`. Rather than specifying overlapping rules, one can also use Curry's *choice* operator "`?`" in order to define non-deterministic operations. The *choice* operator is predefined as follows

```
x ? _ = x
_ ? y = y
```

Apart from non-deterministic operations Curry also enables computations with partial information by using free variables in expressions instead of standard input values.[3]

For instance, the expression "let x free in not x" is reduced to the results {x=False} True and {x=True} False by binding the free variable x appropriately.

Due to the support of non-determinism and partial data structures, Curry uses an alternative evaluation mechanism compared to Haskell, namely needed narrowing [3]. Basically, needed narrowing corresponds to lazy evaluation using unification instead of pattern matching for the passing of parameters. In case an argument of a function which is required for further evaluation contains a free variable, this variable is bound to a constructor term so that evaluation can continue.

Furthermore, there is a core language of Curry named FlatCurry which provides a simplified representation of programs. Due to its simplicity, it is common practice to implement analysis tools and transformations for FlatCurry rather than for full Curry. An abstract representation of the syntax of FlatCurry programs is depicted in Fig. 1 where sequences of objects $o_1, \ldots, o_n$ are denoted by $\overline{o_n}$. For the sake of simplicity, we assume in the following that literals and literal pattern like numbers or characters are represented as nullary constructors and constructor pattern, respectively.

$$
\begin{array}{llll}
P & ::= & \overline{D_m} & \text{(program)} \\
D & ::= & f(\overline{x_n}) = e & \text{(defined function)} \\
e & ::= & x & \text{(variable)} \\
  & | & c(\overline{e_k}) & \text{(constructor call)} \\
  & | & f(\overline{e_k}) & \text{(function call)} \\
  & | & \text{let } \{ \overline{x_n = e_n} \} \text{ in } e & \text{(recursive let binding)} \\
  & | & \text{let } \overline{x_n} \text{ free in } e & \text{(free variables)} \\
  & | & e_1 \ ? \ e_2 & \text{(non-deterministic choice)} \\
  & | & \text{case}_{id} \ e \text{ of } \{ \overline{p_k \rightarrow e_k} \} & \text{(case expression, } p_i \text{ pairwise different)} \\
p & ::= & c(\overline{x_n}) & \text{(constructor pattern)}
\end{array}
$$

Fig. 1. The FlatCurry representation of programs

A FlatCurry program consists of a sequence of function definitions. Every function is specified by a single rule consisting of pairwise different variables $\overline{x_n}$ on its left-hand side and an expression on its right-hand side.[4] Any pattern matching in the original Curry program has been made explicit by the use of case expressions with pairwise distinct constructor patterns.

All case expressions include a unique identifier id. Additionally, all local function declarations have been lifted to the top level in FlatCurry.

---

[3] Note that variables need to be explicitly declared as free.

[4] Note that higher-order applications are represented in FlatCurry using a predefined operator named apply.

FlatCurry is not only the basis for the implementation of analysis tools but also for the description of Curry's semantics. In Sect. 4 we consider the operational semantics of FlatCurry which was originally presented in [1], revised by Hanus and Peemöller in [12] and augmented by us for concolic testing.

3   Satisfiability Modulo Theories

In this section we give a brief overview of satisfiability modulo theories (SMT) and its dedicated solvers. Moreover, we present SMT-LIB, a library providing common standards and benchmarks for the comparison of SMT solvers.

3.1   General Overview

An SMT problem [7] is a decision problem which can be represented as first-order logic formulas containing special predicate symbols with additional interpretations. These interpretations are predefined by so called theories which can be applied during modelling and solving of SMT problems. For instance, there are theories for integer and real arithmetic, but also for uninterpreted functions, arrays, bit-vectors and recursive datatypes. Hence, an SMT instance is a generalization of a Boolean satisfiability (SAT) instance including additional predicates from various underlying theories.

There is a wide range of applications for SMT solving, for example software verification, constraint solving, planning and software testing - to mention only a few. There are also many SMT solvers implementing various APIs and providing different built-in theories. In this work, we focus on the Z3[5] solver developed by Microsoft [6]. Z3 is an efficient, open-source SMT solver supporting the SMT-LIB standard. We primarily chose Z3, because, in addition to basic types like integers, it provides a theory for the definition of algebraic data types which have just recently been added to the SMT-LIB standard.

3.2   SMT-LIB

As mentioned above, SMT-LIB[6] is a library which aims at facilitating research in the SMT sector. Among other things, it provides descriptions of background theories, benchmarks for the comparison of SMT solvers, as well as a standardized input and output language for such solvers [4]. When we refer to SMT-LIB in the following, this input and output language is meant.

An SMT-LIB script is a sequence of commands describing an SMT problem. For instance, the `declare-const` command declares a constant of given type (respectively sort). Z3 internally maintains a stack of declarations and formulas provided by the user. In order to add a formula to this stack, we can use the `assert` command. As mentioned before, a formula is a first-order formula

---

[5] https://github.com/Z3Prover/z3.

[6] http://smtlib.cs.uiowa.edu/index.shtml.

including predicate symbols like < or + with additional interpretations. With the command **check-sat** we can ask the solver to check the satisfiability of the current formulas on the stack. If the formulas are satisfiable, Z3 will answer with **sat**, otherwise with **unsat**. In case Z3 can not determine the satisfiability of a formula, it will return **unknown**. If a formula is satisfiable, i.e., there is an interpretation for the user-declared constants, which makes the asserted formulas true, then we can retrieve the whole interpretation or only single bindings using the commands **get-model** and **get-value**, respectively.

In addition to these commands, Z3 also supports the declaration of polymorphic algebraic data types via the command **declare-datatype**.[7] After their declaration, the type and value constructors can be used like any predefined sort or value.

We conclude this section with a small SMT-LIB script demonstrating some of the commands above. Reconsidering the example from the introduction with the initial call "nthElem [42] 0", we demonstrate the representation of path constraints in SMT-LIB. As mentioned above, during concolic execution we do not only consider the concrete call but also a symbolic one, namely "nthElem xs n". While evaluating the given expression, the variables from the symbolic call are constrained by the branch decisions made along the concrete execution path. The path constraints for the given call can be represented by the formula $xs = y : ys \land n = 0$, where xs, y, ys and n are symbolic variables and : is the constructor for non-empty lists. In order to compute input data which drives the evaluation along an alternative execution path, we can negate a particular path constraint and try to solve the resulting formula with the SMT solver. For instance, we can negate the first constraint of the example above and represent the resulting formula in SMT-LIB as follows.

```
1 (declare-datatype List (par (A) ((nil)
2                                 (cons (head A) (tail (List A)))))))
3 (declare-const xs (List Int))
4 (declare-const n Int)
5 (assert (and (forall ((y Int) (ys (List Int))) (not (= xs (cons y ys))))
6              (= n 0)))
```

List. 2. Representation of path constraints in SMT-LIB

The first two lines show the declaration of a type representing polymorphic lists in SMT-LIB. In contrast to Curry, data type declarations in SMT-LIB also include selector definitions for the arguments of constructors like **head**. Lines 3 and 4 include the necessary constant declarations for the model. The SMT-LIB formula representing the negated path constraint is depicted in lines 5 and 6. Note that we need to universally quantify the arguments of the **cons** constructor in the formula in order to receive an alternative constructor binding for **xs**. Otherwise, Z3 will just bind **xs** to a non-empty list with more elements.

---

[7] Note that we present the syntax of the **declare-datatype** command as it is specified by the SMT-LIB standard version 2.6. At the moment of writing, version 2.6 had just been released and Z3 still used an alternative syntax for the declaration of data types.

If we ask Z3 to check the satisfiability of this problem and return a binding for xs and n, if possible, it will yield sat and the answer ((xs nil) (n 0)).

4   Concolic Interpretation of FlatCurry Code

In this section we describe the concolic interpretation of FlatCurry programs. First, we explain at which points of evaluation symbolic information has to be traced and which information is required for the generation of path constraints. Then, we present an operational semantics for FlatCurry programs which enables the tracing of this information during evaluation.

4.1   Tracing of Symbolic Information

As we have sketched in the introductory example in Sect. 1, the basic idea of concolic testing is to evaluate a program using concrete input data and collect symbolic information at the same time. This symbolic information corresponds to the branch decisions made along a concrete execution path.

Recall that Curry programs use (overlapping) rules and pattern matching for case distinctions. In FlatCurry programs, overlapping rules and pattern matching have been made explicit by the use of non-deterministic choices and case expressions, respectively. As non-deterministic choices and case expressions are the only kinds of branches included in FlatCurry programs, these expressions are the ones of interest for the collection of symbolic information during evaluation.

First, we take a look at purely functional programs. Reconsider the introductory example in List. 1. In the following listing the FlatCurry representation of nthElem is depicted. Note that guards are transformed to case expressions with the respective conditions as arguments.

```
nthElem xs n = case₁ xs of
    []   → Nothing
    y:ys → case₂ n == 0 of
            True  → Just y
            False → case₃ n > 0 of True  → nthElem ys (n - 1)
                                   False → failed
```

We assume that the concolic execution starts with the call "nthElem [42] 0". The case expressions in the program demand their arguments to be evaluated to head normal form, so that a matching branch can be selected. Thus, first [42] is reduced to head normal form selecting the second branch of $case_1$. Next, the head normal form of n == 0 is computed selecting the first branch of $case_2$ and yielding the result Just 42.

In order to reproduce this specific execution path for nthElem, we maintain a symbolic variable for every branch decision made along this path and store this variable together with the selected constructor. For branch decisions which involve comparison operators on numerical literals, e.g. n == 0, we save the constraint associated with the chosen branch. In addition, we always store the case identifier and the index of the selected branch.[8] This information is needed during search to keep track of already visited branches. Hence, we receive the following symbolic trace for the given example: $[(\texttt{case}_1, 2/2, xs_{sym} , (\texttt{:})), (\texttt{case}_2, 1/2, n_{sym} = 0)]$.

Before we conclude this subsection, we take a look at a non-deterministic program. Below the FlatCurry representation of insertND is shown.

```
insertND x xs = case₁ xs of []   → [x]
                            y:ys → (x : y : ys) ? (y : insertND x ys)
```

We consider the call "insertND True [False]", which evaluates to the non-deterministic results [True,False] and [False,True]. Regarding the collection of symbolic information in such programs, two approaches are possible: On the one hand, we could generate traces including non-deterministic branch decisions. In that case we would receive a trace which selects the second branch of case₁ followed by a non-deterministic choice between an empty trace and one selecting the first branch of case₁ for the example considered above. On the other hand, we could encapsulate any non-determinism by constructing a search tree during evaluation. The non-deterministic choices occuring in the program would correspond to the branches of this tree and its leaves would include the various non-deterministic results as well as the respective symbolic trace. Afterwards, we could explore this search tree collecting all possible traces in a list.

Since the interpreter, on which we based the implementation of *ccti*, already supported encapsulation of non-determinism, we chose the latter approach. Hence, the following list of traces is computed for the given example:

```
[ [(case₁, 2/2, xs_{sym₁}, (:))]
, [(case₁, 2/2, xs_{sym₁}, (:)),(case₁, 1/2, xs_{sym₂}, [])] ]
```

For programs including free variables in case expressions we apply narrowing during evaluation to consider all possible bindings for these variables and trace the branch decisions resulting from these bindings accordingly. Note that all traces resulting from a non-deterministic computation or a narrowed free variable are considered during the search for alternative execution paths.

4.2   Augmented Semantics for Concolic Execution

We conclude this section with a description of an augmented operational semantics for FlatCurry enabling the collection of symbolic information during evaluation.

The semantics for concolic execution presented below addresses normalized FlatCurry. During normalization of a FlatCurry program, constructor and function calls are flattened as well as case expressions. For this purpose, we introduce let bindings for the arguments of calls and case expressions, e.g. the function call "not False" is flattened to the expression "let x1 = False in not x1" where x1 is a fresh variable.

---

[8] Note that in the actual implementation further information is collected which is required for the transformation of FlatCurry to SMT-LIB and vice versa.

We use the following definitions and notations in the presentation of the semantics for concolic execution of normalized FlatCurry programs:

1. $\mathcal{V}$ is a set of variables.
2. Exp is a set of FlatCurry expressions.
3. The symbol "free" denotes a free variable.
4. A heap is a partial mapping from variables to either FlatCurry expressions or to the special symbol "free": $Heap = \mathcal{V} \rightarrow \{\text{free}\} \uplus Exp$
5. The empty heap is denoted by $[]$.
6. $\Gamma[x]$ represents the value a variable x is bound to in a heap Γ.
7. $\Gamma[x \mapsto e]$ corresponds to a heap Γ' with $\Gamma'[x] = e$ and $\Gamma'[y] = \Gamma[y]$ for all $y \neq x$.
8. A value is either a free variable which is bound in the associated heap or a constructor applied to a sequence of variables: $Value ::= x \mid c(\overline{x_n})$
9. A symbolic trace T is a list of $SymInfo$ objects.
10. A $SymInfo$ object is a tuple consisting of a case identifier, a branch number, a symbolic variable and the identifier of a FlatCurry constructor.
11. The operation $++$ concatenates two lists.

The operational semantics (also referred to as the natural semantics) of normalized FlatCurry uses a heap structure to represent the sharing of expressions and computes a value for a given FlatCurry expression. In addition to this structure we use a symbolic trace to collect and pass symbolic information during evaluation. This trace is extended whenever a branch decision has been made.

The individual evaluation steps of the natural semantics are formalized using the inference rules depicted in Fig. 2. The inference rules of the semantics include judgements of the form $\Gamma, T : e \Downarrow \Delta, \Upsilon : v$ which can be read as "the FlatCurry expression e under the heap Γ and with incoming symbolic trace T evaluates to value v, the (possibly modified) heap Δ and the (possibly extended) trace Υ".

Apart from the rules (Select) and (Guess), the augmented semantics is equivalent with the FlatCurry semantics presented in [12] except for the fact that symbolic traces are additionally passed through the judgements. Below we give a short description for every rule and explain the modifications to the rules (Select) and (Guess) which are required for concolic execution.

(Value) A value can not be further evaluated and, thus, is directly returned.
(VarExp) If a variable which is bound to an expression in the current heap is evaluated, the associated expression is evaluated to a value and returned. Furthermore, the heap is updated correspondingly to enable the sharing of subexpressions.
(Fun) Flattened function calls are further evaluated by evaluating the right-hand side of the function. For that reason, we assume that the program P is a global parameter of the calculus. In order to prevent name clashes, we apply a renaming substitution σ whenever new variables are introduced during evaluation.

(Let) The bindings of a `let` expression are renamed and then added to the heap. After that, the main expression of the `let` e is evaluated with respect to the bindings.

(Or) For the evaluation of non-deterministic choices one subexpression is chosen non-deterministically to be further evaluated.

(Free) Similar to ordinary `let` expressions logic variables are renamed and added to the heap. Then, the evaluation continues with the main expression e.

(Select) In case the inspected expression of a `case` expression is reducible to a constructor-rooted term, the right-hand side of the corresponding `case` alternative is selected and further evaluated. In addition the trace is extended with symbolic information binding the symbolic variable associated with the case expression to the chosen constructor.

(Guess) If the argument of a `case` expression evaluates to a free variable. One of the `case` alternatives is non-deterministically chosen. The free variable is bound to the corresponding pattern and any variables inside this pattern are bound as free. Moreover, depending on the selected alternative the symbolic variable associated with the case expression is bound to the respective constructor.

(Value) $\quad \Gamma, T : v \Downarrow \Gamma, T : v \quad$ where $v = c(\overline{x_n})$ or $v \in \mathcal{V}$ with $\Gamma[v] = \text{free}$

(VarExp) $\dfrac{\Gamma, T : e \Downarrow \Delta, \Upsilon : v}{\Gamma[x \mapsto e], T : x \Downarrow \Delta[x \mapsto v], \Upsilon : v} \quad$ where $e \notin \{\text{free}\}$

(Fun) $\dfrac{\Gamma, T : \sigma(e) \Downarrow \Delta, \Upsilon : v}{\Gamma, T : f(\overline{y_n}) \Downarrow \Delta, \Upsilon : v} \quad$ where $f(\overline{x_n}) = e \in P$, $\sigma = \{\overline{x_n \mapsto y_n}\}$

(Let) $\dfrac{\Gamma[\overline{y_k \mapsto \sigma(e_k)}], T : \sigma(e) \Downarrow \Delta, \Upsilon : v}{\Gamma, T : \texttt{let } \{ \overline{x_k = e_k} \} \texttt{ in } e \Downarrow \Delta, \Upsilon : v} \quad$ where $\sigma = \{\overline{x_k \mapsto y_k}\}, \overline{y_k}$ fresh

(Or) $\dfrac{\Gamma, T : e_i \Downarrow \Delta, \Upsilon : v}{\Gamma, T : e_1 ? e_2 \Downarrow \Delta, \Upsilon : v} \quad$ where $i \in \{1, 2\}$

(Free) $\dfrac{\Gamma[\overline{y_n \mapsto \text{free}}], T : \sigma(e) \Downarrow \Delta, \Upsilon : v}{\Gamma, T : \texttt{let } \overline{x_n} \texttt{ free in } e \Downarrow \Delta, \Upsilon : v} \quad$ where $\sigma = \{\overline{x_n \mapsto y_n}\}, \overline{y_n}$ fresh

(Select) $\dfrac{\Gamma, T : x \Downarrow \Delta, \Upsilon : c(\overline{y_n}) \qquad \Delta, \Phi : \sigma(e_i) \Downarrow \Theta, X : v}{\Gamma, T : \texttt{case}_{id} \ x \texttt{ of } \{ \overline{p_k \to e_k} \} \Downarrow \Theta, X : v}$

$\quad$ where $p_i = c(\overline{x_n}), \Phi = \Upsilon + [(id, i/k, x, c)], \sigma = \{\overline{x_n \mapsto y_n}\}$

(Guess) $\dfrac{\Gamma, T : x \Downarrow \Delta[y \mapsto \text{free}], \Upsilon : y \qquad \Theta, \Phi : \sigma(e_i) \Downarrow \Lambda, X : v}{\Gamma, T : \texttt{case}_{id} \ x \texttt{ of } \{ \overline{p_k \to e_k} \} \Downarrow \Lambda, X : v}$

$\quad$ where $i \in \{1, \ldots, k\}, p_i = c(\overline{x_n}), \Theta = \Delta[y \mapsto \sigma(p_i), \overline{y_n \mapsto \text{free}}],$
$\quad \Phi = \Upsilon + [(id, i/k, x, c)], \sigma = \{\overline{x_n \mapsto y_n}\}, \overline{y_n}$ fresh

Fig. 2. Natural semantics for concolic execution of normalized FlatCurry programs

5 Search Strategy

In the previous section we described the collection of symbolic traces during the evaluation of a FlatCurry expression. A single trace corresponds to a path through the associated symbolic execution tree and the symbolic information derived from a single case expression corresponds to a node of this tree. To produce new test cases, we have to search for unexplored paths through that tree. Hence, we need to select a node with unvisited branches and negate the path constraint associated with that node. If there is a solution for the resulting constraints, new input data which will drive execution along one of these branches can be computed.

A naive strategy - similar to the one presented in [9] - is to choose the first node with unvisited branches which is closest to the root of the symbolic execution tree. This strategy which basically corresponds to a breadth-first-like search with branch coverage is currently used in *ccti*.

The general search algorithm of *ccti* is depicted in Fig. 3. Basically, two data structures are used during search: On the one hand, there is a priority queue Q storing the nodes of the symbolic execution tree in a strategy-defined order. The naive strategy mentioned above can be implemented by using a priority function preferring the node with the lowest depth in the tree. On the other hand, we maintain a map of all case expressions M including still unvisited branches. Note that we provide a generic interface for the implementation of the search so that both data structures can be easily replaced to implement alternative search strategies and coverage criteria.

The central function of the search is SEARCHLOOP. It evaluates the function to be tested f with the given inputs *in* yielding potentially non-deterministic results *res* and a list of symbolic traces $T_s s$. The inputs and the results form a new test case which is added to the set of test cases T. By calling PROCESS the traces are processed to update the priority queue and the case map with the information collected during the previous evaluation. More precisely, for every *SymInfo* object (cid, bnr, x, c) included in a trace, the queue is extended with information on the case identifier *cid*, the associated symbolic variable x and the set of path constraints C using ENQUEUE. This set includes all constraints associated with the path leading from the root of the symbolic execution tree up to that particular node. VISIT marks the selected branch *bnr* as visited and adds the chosen FlatCurry constructor c to the set of known constructors for a particular *cid*. Before processing further *SymInfo* objects of the trace, the set of path constraints is extended with the constructor decision made in the current *SymInfo* object by applying CONSTR.

While the priority queue Q is not empty, we dequeue the next entry (d, cid, x, C) from the queue. In order to compute input data driving execution along an alternative branch of the case expression identified by *cid*, we have to generate an appropriate path constraint. Hence, we select the set of already known constructors for that case expression from the case map and constrain the associated symbolic variable x to be different than any known constructor by calling GETCONS and NONEOF, respectively. Next, we extend the set of path constraints with the new constraint and call SOLVE to apply the SMT solver. In case the constraints are satisfiable, the resulting model is translated into valid FlatCurry inputs with TOFCY and a new iteration of the concolic search is started. In case the constraints are not satisfiable or the solver can not determine their satisfiability, we proceed with the search considering constraints along a different path.

SEARCHLOOP f $args$ T Q M =
  let $(res, T_s s)$ = EVAL $(f\ args)$
    T' = $T \cup \{(args, res)\}$
    (Q', M') = **fold** PROCESS $(Q, M)\ T_s s$
  in SEARCHINPUT $f\ T'\ Q'\ M'$

PROCESS $[\,]$ Q M C d = (Q, M)
PROCESS $((cid, bnr, x, c) : T_s)$ Q M C d =
  let Q' = ENQUEUE Q (d, cid, x, C)
    M' = VISIT M (cid, bnr, c)
    C' = $C \cup \{$CONSTR x $c\}$
  in PROCESS T_s Q' M' C' $(d + 1)$

SEARCHINPUT $f\ T\ \varnothing$ $M = T$
SEARCHINPUT $f\ T$ $((d, cid, x, C) :> Q')$ M =
  let cs = GETCONS M cid
    pc = NONEOF x cs
    sa = SOLVE $(C \cup \{pc\})$
  in **if** $sa == (sat, m)$
    then SEARCHLOOP f (TOFCY m) T Q' M
    else SEARCHINPUT $f\ T\ Q'\ M$

C	set of path constraints
M	map of case expressions
Q	priority queue
T	set of test cases
$T_s s$	list of symbolic traces
T_s	symbolic trace
$args$	FlatCurry arguments of f
bnr	branch number
cid	case identifier
c	FlatCurry constructor
cs	already known constructors
d	depth
m	SMT model
pc	new path constraint
res	FlatCurry results of f
sa	answer from SMT solver
x	symbolic variable

Fig. 3. Basic search algorithm of *ccti*

6 Implementation

This section gives a brief overview of the implementation of *ccti* which is completely implemented in Curry. The concolic execution part of *ccti* is implemented by a FlatCurry interpreter implementing the augmented operational semantics of FlatCurry presented in Sect. 4.2. In this section we focus on the integration of SMT in Curry.

As explained before, we want to apply SMT solvers to compute alternative inputs from the information included in a symbolic trace. Hence, we need to transform the path constraints, i.e., the constructor decisions made along an execution path, into an SMT-LIB formula.

In Sect. 3.2 we demonstrated how to model path constraints for our running example in SMT-LIB (see List. 2) by declaring corresponding SMT-LIB types for the types used in the considered Curry program and representing path constraints as simple relational formulas on constructor terms and integers, respectively.

To simplify the translation of path constraints we provide some libraries in Curry. Among these are a representation of SMT-LIB scripts as abstract data types, a pretty printer and a parser to send String representations of the scripts to the SMT solver and parse its responses. Moreover, an interface to call SMT solvers via Curry as well as a transformation library to convert FlatCurry expressions to SMT-LIB terms and vice versa are provided.

When we run *ccti* on a Curry module, the module and all its dependent modules are parsed to FlatCurry.[9] To prepare for a type-safe translation of FlatCurry constructor calls to corresponding SMT-LIB terms, the transformation library then builds up bidirectional maps mapping both FlatCurry type and value constructors to their associated sort or term on SMT-LIB side. Furthermore, a corresponding SMT-LIB declaration for all data types occuring in the program is generated.

We also construct a type environment mapping the symbolic variables occuring in the trace to their FlatCurry type and SMT-LIB sort, respectively. On the one hand, this information is required for the declaration of variables in SMT-LIB. On the other hand, it is needed to transform possible results found by the solver into type correct FlatCurry expressions. Note that these results correspond to alternative input data for the function to be tested and, thus, their FlatCurry representation is required to start the next iteration of the concolic execution.

We conclude this section taking a look at another example for the generation of an SMT-LIB script. Reconsidering our running example let us assume that we call *ccti* with the initial call "nthElem [] 0" this time. The concolic execution of this call produces a symbolic trace which constrains the list argument to be an empty list. If we try to model this path constraint in SMT-LIB, there is a problem: nthElem is defined on polymorphic lists and the example call does not specify a type for the list elements. Nevertheless, that type information has to be determined for the translation, because SMT-LIB does not allow the declaration of polymorphic constants. Hence, during translation of FlatCurry types to SMT-LIB sorts, all occurrences of type variables are instantiated with a monomorphic type. With regard to the generation of test cases, it seems reasonable to use a type for instantiation which includes more than one value but is also simple. For this reason, we use Curry's Ordering type, which is equivalent to the one in Haskell, whenever polymorphic types need to be instantiated.

Reconsidering the example call of nthElem from above, the resulting path constraint is represented in SMT-LIB as shown below and running Z3 with this script yields the answer ((xs (cons lt nil))).

---

[9] Note that we actually use a variant of FlatCurry called TypedFlatCurry which corresponds to FlatCurry with the exception that expressions are additionally annotated with type information.

```
(declare-datatype Ordering ( (lt) (eq) (gt) )
(declare-datatype List (par (A) ((nil)
                                 (cons (head A) (tail (List A)))))))
(declare-const xs (List Ordering))
(assert (not (= xs nil)))
```

7   Application of *ccti*

In this section we want to take a closer look at the usage and applicability of *ccti* regarding some practical examples. Note that this work is still in progress. Hence, we only discuss the general applicability of our approach.

Currently, *ccti* expects a Curry module to include a main function calling the function to be tested with user-specified inputs in order to initiate the concolic execution. For the future, we plan to support concolic testing of multiple functions in a single run of *ccti* as well as a random-based generation of initial input data.

For the following examples we applied the search algorithm presented in Sect. 5 using branch coverage, i.e. execute every program branch at least once. Applying *ccti* to our running example with the initial call "nthElem [42] 0" produces four test cases. Among these is also one resulting in a failure, because our implementation of nthElem does no handle negative indices. The minimum number of three test cases (including the one with the failure) is generated, if we call *ccti* with a list with at least two elements and an index greater than 0.

```
data Nat = IHi | O Nat | I Nat

add IHi    y     = succ y
add (O x)  IHi   = I x
add (O x)  (O y) = O (add x y)
add (O x)  (I y) = I (add x y)
add (I x)  IHi   = O (succ x)
add (I x)  (O y) = I (add x y)
add (I x)  (I y) = O (add (succ x) y)

succ IHi   = O IHi
succ (O x) = I x
succ (I x) = O (succ x)
```

List. 3. Addition of binary numbers

For a more complex example, we consider the implementation of an addition operation on a representation of binary numbers in Curry, which is depicted in List. 3. Calling *ccti* with the initial call "add IHi IHi", generates nine test cases. These test cases cover all branches, but, in fact, even six test cases would be sufficient for full coverage. This minimum number of test cases is found by *ccti*, if we use "add IHi (I (O IHi))" to initialize the concolic execution.

We conclude this brief case study considering our running example for a non-deterministic operation, namely `insertND`. Calling *ccti* with "`insertND True []`", the following two test cases will be produced.[10]

```
insertND True []      = {[True]}
insertND True [False] = {[True, False], [False, True]}
```

Finally, we take a look at an example showing a limitation of branch coverage. List. 4 shows a definition of an operation to compute all permutations of a list.

```
perm []     = []
perm (x:xs) = insertND x (perm xs)
```
List. 4. Permutation of lists

Considering the initial call "`perm [False]`" we receive the single test case `perm [False] = [False]`. This test case covers both rules of `perm` but only the first rule of `insertND`. Branch coverage makes no difference between different calls of a function. A branch is already covered when an arbitrary call of a function selects that particular branch. Thus, in our example, it is sufficient, if the non-empty list branch of `perm` is visited in either of the two calls, i.e. the top-level call of `perm` or the recursive one. Since our initial call already covers the non-empty list branch when the top-level call of `perm` is evaluated, *ccti* does not consider this branch for the recursive call, when using branch coverage. For that reason, the recursive call of `perm` can only result in an empty list, and thus the second and third rule of `insertND` are never tested.

For full coverage of `insertND`, *ccti* needs to produce an input list for `perm` with at least two elements. To compute such a list, we need to reconsider all branches in the recursive call of `perm`, even if they already have been visited in the top-level call, i.e. different calls of the same function have to be covered, separately. This approach corresponds to the *function coverage* criterion discussed in [8]. Using a prototypical implementation of function coverage with *ccti* yields three additional test cases, a redundant one, one with an empty input list and one with a two elemented input list. Hence, with function coverage `perm` and `insertND` are fully covered.

8 Conclusions and Future Work

In this work, we have presented *ccti*, a tool for concolic testing of Curry programs. We have extended the operational semantics of FlatCurry - a simplified core language of Curry - to collect the necessary information for concolic testing during evaluations. *ccti* is based on a FlatCurry interpreter implementing this semantics. Applying an external SMT solver integrated in Curry, we compute input data for the exploration of alternative execution paths. In contrast to the narrowing-based test case generation, path constraints on literals can be mapped directly to SMT by using suitable theories of the solver.

---

[10] Note that we use a set notation to represent multiple non-deterministic results.

ccti provides support for the implementation of alternative search strategies and coverage criteria. Although the achievement of full program coverage highly depends on the coverage criterion, first applications of *ccti* show that our approach is applicable for the automated generation of test cases of functional logic programs.

For future work, we plan to further evaluate the applicability of *ccti* when using different strategies and coverage criteria to perform concolic execution on more complex programs. Another interesting aspect might be to transfer the ideas from this interpreter-based approach to an instrumentation-based one. This might be possible by instrumentalizing TypedFlatCurry programs before compilation.

References

1. Albert, E., Hanus, M., Huch, F., Oliver, J., Vidal, G.: Operational semantics for declarative multi-paradigm languages. J. Symbolic Comput. **40**(1), 795–829 (2005)
2. Amaral, C., Florido, M., Santos Costa, V.: PrologCheck – property-based testing in prolog. In: Codish, M., Sumii, E. (eds.) FLOPS 2014. LNCS, vol. 8475, pp. 1–17. Springer, Cham (2014). https://doi.org/10.1007/978-3-319-07151-0_1
3. Antoy, S., Echahed, R., Hanus, M.: A needed narrowing strategy. J. ACM **47**(4), 776–822 (2000)
4. Barrett, C., Fontaine, P., Tinelli, C.: The SMT-LIB standard: Version 2.5. Technical report, Department of Computer Science, The University of Iowa (2015). www.SMT-LIB.org
5. Claessen, K., Hughes, J.: QuickCheck: a lightweight tool for random testing of Haskell programs. In: International Conference on Functional Programming (ICFP 2000), pp. 268–279. ACM Press (2000)
6. de Moura, L., Bjørner, N.: Z3: an efficient SMT solver. In: Ramakrishnan, C.R., Rehof, J. (eds.) TACAS 2008. LNCS, vol. 4963, pp. 337–340. Springer, Heidelberg (2008). https://doi.org/10.1007/978-3-540-78800-3_24
7. de Moura, L., Dutertre, B., Shankar, N.: A tutorial on satisfiability modulo theories. In: Damm, W., Hermanns, H. (eds.) CAV 2007. LNCS, vol. 4590, pp. 20–36. Springer, Heidelberg (2007). https://doi.org/10.1007/978-3-540-73368-3_5
8. Fischer, S., Kuchen, H.: Systematic generation of glass-box test cases for functional logic programs. In: Proceedings of the 9th ACM SIGPLAN International Conference on Principles and Practice of Declarative Programming (PPDP 2007), pp. 63–74. ACM Press (2007)
9. Giantsios, A., Papaspyrou, N.S., Sagonas, K.F.: Concolic testing for functional languages. In: Proceedings of the 17th International Symposium on Principles and Practice of Declarative Programming, 14–16 July 2015, Siena, Italy (2015)
10. Godefroid, P., Klarlund, N., Sen, K.: DART: directed automated random testing. In: Proceedings of the ACM SIGPLAN 2005 Conference on Programming Language Design and Implementation, 12–15 June 2005, Chicago, IL, USA (2005)
11. Hanus, M.: CurryCheck: checking properties of curry programs. In: Hermenegildo, M.V., Lopez-Garcia, P. (eds.) LOPSTR 2016. LNCS, vol. 10184, pp. 222–239. Springer, Cham (2017). https://doi.org/10.1007/978-3-319-63139-4_13

12. Hanus, M., Peemöller, B.: A partial evaluator for Curry. In: Proceedings of the 28th Workshop on (Constraint) Logic Programming (WLP 2014) Proceedings of the 23rd International Workshop on Functional and (Constraint) Logic Programming, 15–17 September 2014, Wittenberg, Germany (2014)

13. Hanus, M. (ed.): Curry: An Integrated Functional Logic Language (vers. 0.9.0) (2016). http://www.curry-language.org

14. Mesnard, F., Payet, E., Vidal, G.: Concolic testing in logic programming. TPLP **15**(4–5) (2015)

15. Mesnard, F., Payet, É., Vidal, G.: On the completeness of selective unification in concolic testing of logic programs. In: Hermenegildo, M.V., Lopez-Garcia, P. (eds.) LOPSTR 2016. LNCS, vol. 10184, pp. 205–221. Springer, Cham (2017). https://doi.org/10.1007/978-3-319-63139-4_12

16. Palacios, A., Vidal, G.: Concolic execution in functional programming by program instrumentation. In: Falaschi, M. (ed.) LOPSTR 2015. LNCS, vol. 9527, pp. 277–292. Springer, Cham (2015). https://doi.org/10.1007/978-3-319-27436-2_17

17. Sen, K., Agha, G.: CUTE and jCUTE: concolic unit testing and explicit path model-checking tools. In: Ball, T., Jones, R.B. (eds.) CAV 2006. LNCS, vol. 4144, pp. 419–423. Springer, Heidelberg (2006). https://doi.org/10.1007/11817963_38

18. Sen, K., Marinov, D., Agha, G.: Cute: a concolic unit testing engine for C. In: Proceedings of the 10th European Software Engineering Conference Held Jointly with 13th ACM SIGSOFT International Symposium on Foundations of Software Engineering. ACM (2005)

Declarative XML Schema Validation with SWI–Prolog
System Description

Falco Nogatz[✉] and Jona Kalkus

Department of Computer Science, University of Würzburg, Am Hubland,
97074 Würzburg, Germany
`falco.nogatz@uni-wuerzburg.de,`
`jona.kalkus@stud-mail.uni-wuerzburg.de`

Abstract. XML Schema is a well–established mechanism to define the structure and constrain the content of an XML document. While this approach taken by itself is declarative, currently available tools for XML validation are not. In this paper we introduce an implementation of an XSD validator in SWI–Prolog, made publicly available as the package *library(xsd)*. Our approach is based on flattening the XSD and XML documents into Prolog facts. The top–down validation makes great use of Prolog's backtracking and unification capabilities. To ensure the compliance to the XSD standard and to support the test–driven development, we have created a test framework based on the Test Anything Protocol and SWI–Prolog's quasi–quotations.

Keywords: XML Schema · XSD · XML · SWI–Prolog · Validation
Quasi–quotation

1 Introduction

The *Extensible Markup Language* (XML) [1] is one of the most used data formats to store and exchange structured data. Especially in the context of web services, XML documents are often used for data transfer and as configuration files. These use cases emphasise the importance for tools that ensure an expected format of the used XML documents.

One approach to specify the structure and content of XML documents is to use an *XML Schema Definition* (XSD) [2]. It is used to specify the allowed elements in an XML document, their data types, and additional rules the document has to comply with. While every XML document has to be *well–formed*, i.e. it has to follow the general syntax rules for XML, XSD is used to ensure *validity* in terms of conformity according to the specified data types and rules.

Version 1.0 of the XSD specification was originally published in 2001, a second edition followed in 2004. Since then, a great number of tools to validate XML documents against a given XSD has been published. In 2012, the XSD 1.1

© Springer Nature Switzerland AG 2018
D. Seipel et al. (Eds.): DECLARE 2017, LNAI 10997, pp. 187–197, 2018.
https://doi.org/10.1007/978-3-030-00801-7_12

specification [3] became a W3C Recommendation. It introduces new, significant features like the ability to define assertions based on XPATH [4] expressions and conditional type assignments. Although completely backward compatible, these new features require the handling of expressive, declarative rules which can often not be easily added to existing tools, because they are mostly based on imperative programming languages. Therefore, the number of XSD validators which support the most recent XSD 1.1 standard is still limited. Three of the most popular tools with support for XSD 1.1 are: Apache Xerces2 Java[1], Oxygen XML Editor[2], and Saxon XSLT[3].

SWI–Prolog [5] already has good support for XML. Together with Prolog's built–in backtracking and unification abilities, this makes it a good target platform for a new, extensible XSD validation software. In this paper, we present an approach to process XML and XSD files using SWI–Prolog. The validation module *library(xsd)* unfolds a given XML and its XSD into a knowledge base representing the documents as Prolog facts. This way it is possible to define declarative Prolog rules that ensure the schema properties for all instance nodes that can be unified, resulting in a validation where the XML nodes are processed in a top–down manner.

To ensure the compliance to the XSD standard, our implementation comes with a test framework based on the *Test Anything Protocol* (TAP) [6]. It makes great use of SWI–Prolog's quasi–quotations [7] to directly embed example XML documents into Prolog source code as an external domain–specific language.

Our XSD validator is available as a package for SWI–Prolog and listed in its package list at http://www.swi-prolog.org/pack/list?p=xsd. It can be easily installed using **pack_install(xsd)** and used similar to built–in libraries by calling **use_module(library(xsd))**. The validator is published under MIT License as open source at https://github.com/jonakalkus/xsd.

The remainder of this paper is organised as follows. In Sect. 2 we introduce the work with XSD and XML files in SWI–Prolog and present possible representations in Prolog. In Sect. 3, the validation process is described. The embedding of XML into SWI–Prolog using quasi–quotations is presented along with the test framework in Sect. 4. Finally, we conclude with a summary and discussion of future work in Sect. 5.

2   On the Integration of XML in SWI–Prolog

Prolog is well–known for processing natural language. However, Prolog is also an excellent language to work with data given in a formal language. SWI–Prolog is already widely used to process XML documents. Recently, the integration with semantic data given as RDF/XML [8] in Prolog has been an emerging field of research, resulting in great support for RDF/XML in SWI–Prolog [9,10] and

---

[1] http://xerces.apache.org/ [accessed 12 March 2018], Apache License 2.0.

[2] https://www.oxygenxml.com/ [accessed 12 March 2018], proprietary license.

[3] http://www.saxonica.com/ [accessed 12 March 2018], XSD 1.1 support in Saxon Enterprise Edition 9.5, proprietary license.

decent semantic web frameworks like Cliopatria [11]. Their success depends on stable and fast RDF/XML parsers.

2.1   A Motivating Example

As a motivating example, we will consider a small XSD, as shown in Fig. 1. Following the formal description of the XSD language [2], it mainly consists of descriptions of elements, simple and complex types, and attributes. We assume basic knowledge about XSD here and provide only a short, informal description. The XSD characterises XML documents with a single root node <person>, and <name> and <email> child nodes. The example document given in Fig. 2 is valid against this XSD, while the second XML given in Fig. 3 is not valid because of its missing <email> node and the wrong value abc for the attribute no of the data type id.

```
<xs:schema
 xmlns:xs="http://www.w3.org/2001/XMLSchema">
 <xs:element name="person"><xs:complexType>
  <xs:sequence>
   <xs:element name="name" type="xs:string"/>
   <xs:element name="email" type="xs:string"
    maxOccurs="unbounded"/>
  </xs:sequence>
  <xs:attribute name="no" type="id"
  use="required"/>
 </xs:complexType></xs:element>
 <xs:simpleType name="id">
  <xs:restriction base="xs:int">
   <xs:minInclusive value="100"/>
 </xs:restriction></xs:simpleType></xs:schema>
```

Fig. 1. Example XSD

```
<person no="123">
 <name>John Doe</name>
 <email>john@doe.com</email>
 <email>j.doe@example.com</email>
</person>
```

Fig. 2. Valid XML

```
<person no="abc">
 <name>John Doe</name>
</person>
```

Fig. 3. Non–valid XML

The aim of *library(xsd)* is to identify the XML of Fig. 2 as valid, and the XML of Fig. 3 as invalid. It provides a single predicate xsd_validate(+XSD,+XML) which succeeds only for XML documents that are valid according to the given XSD.

2.2   Parsing XML with *library(sgml)*

XML is an application profile of the *Standard Generalized Markup Language* (SGML) [12] and therefore just a subset of SGML. As a result, it is possible to use an SGML parser to load XML files in Prolog. The first SGML parser for SWI–Prolog was created by Anjo Anjewierden and was based on the SP parser[4]. Today's versions of SWI–Prolog come with a faster SGML parser implemented as a C–library [10,13]. Both SGML parsers share the same output format and a

---

[4] http://www.jclark.com/sp/ [accessed 12 March 2018].

similar interface. Since XSD is an application of XML, the SGML parser can be used for both input file formats.

The SGML parser can be used in SWI–Prolog after loading the module *library(sgml)*, which by default is bundled with SWI–Prolog. It provides a predicate `load_structure(+Source,-Out,+Options)` to load structured files like SGML, HTML, or XML. Most importantly we use the options (i) `dialect(xmlns)`, to read in the given files as XML documents using the built–in namespace handling, and (ii) `keep_prefix(true)`, to store the namespace's URI along with the node's type. The latter option requires SWI–Prolog of at least version 7.3.26.

2.3 Nested Term Representations

SWI–Prolog's built–in SGML parser returns a nested list. Each node is represented by a Prolog term of the form

```
element(ns(Prefix,URI):Type,Attributes,Children).
```

For instance, the XSD of Fig. 1 generates the following term:

```
element( ns(xs,'http://www.w3.org/2001/XMLSchema'):schema,
  [ xmlns:xs='http://www.w3.org/2001/XMLSchema' ],
  [ element(
      ns(xs,'http://www.w3.org/2001/XMLSchema'):element,
      [name=seq], [ ... ]), ... ] )
```

Seipel et al. have transformed this data structure into a more convenient form called *field notation* [14]. It is based on association lists and triples of the form

```
Type:Attributes:Children
```

and integrates a declarative query mechanism called FNQUERY [15].

In *library(xsd)*, we use a top–down validation approach, where the validator simultaneously traverses the XSD and XML document, beginning with the `<xs:schema>` resp. root node. At first sight, the nested term representation looks like a good data structure for this approach using tree traversal. However, in XSD it is possible to define types globally (like the simple type `id` in Fig. 1) which are usually referenced by other elements which are not necessarily part of the same XML sub–document. The same applies for named element references using `<xs:element ref="..." />`. Therefore, element types and names would have to be stored globally.

2.4 XML Flattening

The nested term can be avoided by *flattening*: the contained elements are asserted as facts with a unique identifier. Based on the identifier it is possible to retrieve, for example, the parent node, all siblings, or any descendant. In addition to this, globally defined types and named elements can be easily accessed.

In [16], Nogatz et al. introduced *xsd2json*, a tool that translates an XSD into an equivalent JSON Schema using Prolog and Constraint Handling Rules (CHR) [17]. To represent the XSD as CHR constraints, a similar flattening step has been applied. The flattening implemented in *library(xsd)* is for the most part an adapted version, which asserts Prolog facts instead of generating CHR constraints. It can be separately used as `xml_flatten(+XML,?Handle)` in the sub–package *library(xsd/flatten)*. If not provided, it returns a unique identifier `Handle` to reference an already flattened XML file. This handle is part of every asserted fact to distinguish multiple loaded XML files. In the code examples in this paper, we use `xsd` as the handle of a loaded XSD document, and `xml` for the loaded XML document.

The asserted facts are similar[5] to the CHR constraints generated by *xsd2json*:

- `node(Handle, ID, Namespace, Type)`
  For each XML node a new `node/4` predicate is asserted, only holding its `Namespace` and `Type`.
- `node_attribute(Handle, ID, Attribute, Value)`
  For each XML attribute a new `node_attribute/4` is asserted, holding the attribute's name and value.[6]
- `text_node(Handle, ID, Text)`
  If an element's child is simply a text and no nested XML, a `text_node/3` is asserted with its `Text`.

The node's unique identifiers ID are generated inductively: (i) the root node has an ID of `[0]`, and (ii) the ID of all other nodes is of the form `[Position|Parent_ID]`, with `Position` starting from 0 and being incremented for every sibling. This way an element's siblings, ancestors, and descendants can be retrieved by simple unifications based on the element's identifier.

For instance, the flattening of the XSD of Fig. 1 generates the following `node/4` and `node_attribute/4` facts:

```
?- xml_flatten('file.xsd',xsd), listing([node/4,node_attribute/4,text_node/4]).

node(xsd, [      0], ns(xs, 'http://www.w3.org/2001/XMLSchema'), schema).
node(xsd, [   0, 0], ns(xs, 'http://www.w3.org/2001/XMLSchema'), element).
node(xsd, [0, 0, 0], ns(xs, 'http://www.w3.org/2001/XMLSchema'), complexType).
node(xsd, [   1, 0], ns(xs, 'http://www.w3.org/2001/XMLSchema'), simpleType).
node(xsd, [0, 1, 0], ns(xs, 'http://www.w3.org/2001/XMLSchema'), restriction).
% ... and 5 other node/4
node_attribute(xsd, [      0], xmlns:xs, 'http://www.w3.org/2001/XMLSchema').
node_attribute(xsd, [   0, 0], name, person).
node_attribute(xsd, [   1, 0], name, id).
node_attribute(xsd, [0, 1, 0], base, 'xs:int').
% ... and 9 other node_attribute/4
```

---

[5] *xsd2json* is in active use and maintained. It is available as open source at https://github.com/fnogatz/xsd2json (MIT License). Because of recent improvements, the constraint functors have been slightly changed compared to [16].

[6] XML attributes in general do not have any namespaces. For special attributes like the declaration of the namespace prefix `xs` in `xmlns:xs="..."`, this is handled separately.

Because the XSD of Fig. 1 does not contain an XML node with only text content, no `text_node/3` fact is asserted.

3   Top–Down Validation by Simultaneous Tree Traversals

To validate an XML against its XSD both documents are traversed simultaneously. The document's nodes are validated step–by–step, beginning with its root node with the unique identifier of [0], followed by its descendants [0,0], [1,0], and so on. The number of XML and XSD nodes which are involved in a single validation step varies: a single XML element might require several alternative XSD nodes (e.g., in case of `<xs:choice>` definitions); then again several XML elements can be specified by a single XSD node (e.g., in case of `<xs:element maxOccurs="unbounded">`).

We expect that both the XSD and XML are well–formed; the XSD is expected to strictly follow the XSD specification. The main predicate `xsd_validate/2` uses the predicate `validate(+S_Handle,+D_Handle)` of the sub–package *library(xsd/validate)*, which can also be used on its own.

The rules on how to validate a given XML node are stated using the predicate `validate(+D_Handle,+D_ID,?Vals,+S_Handle,+S_ID)`. Given the two current positions in the XML and XSD trees, specified by the appropriate pair (`Handle,ID`), we have implemented rules to confirm its validity. The additional argument `Vals` is an internal counter which is used to ensure, among others, the correct number of elements in a `<xs:sequence>` with respect to the `minOccurs` and `maxOccurs` properties. For common selections we provide predicates like `child(+Handle,?ID,?Child)`, which returns a child of the node with the given ID, and vice versa.

Complex Type Validation. The validation process starts at the root nodes of the XML and XSD document, given by their `D_Handle` resp. `S_Handle`. So `validate/2` is simply implemented as follows:

```
validate(S_Handle, D_Handle) :-
    validate(D_Handle, [0], 1, S_Handle, [0]).
```

library(xsd) has only a single rule that can be applied for this initial goal `validate(xml,[0],1,xsd,[0])`:

```
validate(D_Handle, D_ID, 1, S_Handle, S_ID) :-
    node(S_Handle, S_ID, ns(_, 'http://www.w3.org/2001/XMLSchema'), schema),
    child(S_Handle, S_ID, S_Child),
    node(S_Handle, S_Child, ns(_, 'http://www.w3.org/2001/XMLSchema'), element),
    validate(D_Handle, D_ID, 1, S_Handle, S_Child).
```

It reads as follows: The current XML position is valid if the XSD is a `<xs:schema>` node containing a `<xs:element>` child node which is valid, too. This is correct according to the XSD specification as there can be various root nodes defined in the XSD. They have to be handled as alternatives, i.e. there must be at least one that is valid. If there is one, Prolog's backtracking mechanism will find the appropriate `S_Child` and continues the validation at this point.

Rules that handle the validation of complex types contain `validate/5` predicates in the rule's body to recursively validate all the contained elements. The following more complex source code example demonstrates how to validate a single element:

```
validate(D_Handle, D_ID, Vals, S_Handle, S_ID) :-
    node(S_Handle, S_ID, ns(_, 'http://www.w3.org/2001/XMLSchema'), element),
    attribute(S_Handle, S_ID, minOccurs, Min),
    attribute(S_Handle, S_ID, maxOccurs, Max),
    between(Min, Max, Vals), validate_element(D_Handle, D_ID, S_Handle, S_ID),
    forall(between(Vals+1, Max, Next),
      (get_nth_sibling(D_Handle, D_ID, Next, Next_ID),
         validate(D_Handle, Next_ID, S_Handle, S_ID))).
```

A `<xs:element>` node is only valid if it is allowed at this position according to the `minOccurs` and `maxOccurs` properties set in the XSD's `attribute/4`. In addition, the element must be valid itself, i.e. of the correct type, etc. In the last part, all sibling nodes referenced by the same XSD position are ensured to be valid, too.

The `attribute/4` is a wrapper for the asserted `node_attribute/4` predicate. It takes into account default values according to the XSD specification. If, e.g., a `minOccurs` is set explicitly in the XSD document via `node_attribute/4`, it is returned by `attribute/4`, otherwise the default value of 1 is used.

Simple Type Validation. The leaves of an XML document tree are mostly formed by elements of simple types. *library(xsd)* provides the sub–package *library(xsd/simpletype)*, which validates XSD types like `xs:int`, `xs:string`, etc. It also considers constraining facets like `<xs:minInclusive>`. For pattern–based restrictions as they are used by `<xs:pattern>`, or XSD's `xs:date` and `xs:time` data types, we make use of SWI–Prolog's *library(regex)*[7].

Backtracking. In the tool *xsd2json* as presented in [16], Nogatz et al. also flattened a given XSD in order to translate it into an equivalent JSON Schema. Instead of asserting Prolog facts, the `node`, `node_attribute`, and `text_node` terms are propagated as CHR constraints. The XSD document is later translated using a tree traversal, too. However, the XSD validation in *library(xsd)* makes great use of backtracking which would not be possible in CHR which is a committed–choice language. E.g., when validating elements and sequences with overlapping `minOccurs` and `maxOccurs`, there is often not just a single rule which could be applied. There are also XSD elements which define alternatives explicitly, e.g., in `<xs:choice>`, or the constraining facet `<xs:enumeration>`. These alternatives are directly supported by Prolog's built–in backtracking mechanism.

Performance Improvements Using Memoisation. Prolog's backtracking technique allows a compact definition of the validation rules. However, once the backtracking has to be done, part of the already inferred knowledge gets discarded, even though there are some sub–goals which might occur identically in

---

[7] https://github.com/mndrix/regex [accessed 12 March 2018], The Unlicense.

later computations again. This behaviour can be observed especially for XSD documents with nested `<xs:sequence>` or `<xs:choice>` nodes with high `maxOccurs` properties.

We implemented a wrapper which stores already computed validations in a dynamic predicate `xsd_table(Original_Call,Valid)`. If `validate_tabled/5` is called with arguments that have been checked before, its result `Valid={true,false}` is returned immediately:

```
:- dynamic xsd_table/2.
validate_tabled(D_Handle, D_ID, Vals, S_Handle, S_ID) :-
  ( xsd_table(validate(D_Handle, D_ID, Vals, S_Handle, S_ID), Valid) ->
    !, call(Valid)      % still trigger backtracking if invalid
  ; validate(D_Handle, D_ID, Vals, S_Handle, S_ID) ->
    asserta(xsd_table(validate(D_Handle, D_ID, Vals, S_Handle, S_ID), true))
  ; asserta(xsd_table(validate(D_Handle, D_ID, Vals, S_Handle, S_ID), false)),
    !, false ).         % trigger backtracking
```

This memoisation technique is possible only because an XML fragment is valid against a given XSD fragment independently of its surrounding elements. The triple `(D_ID,Vals,S_ID)` is unique and it is not possible to be valid once and invalid later, or vice versa. In edge cases with many nested `<xs:sequence>` or `<xs:choice>` nodes, this saves up to 98% of the computation time.[8]

Compared to the traditional tabling implementations in Prolog [18,19], this technique also stores *failing* computations. SWI–Prolog's current tabling implementation only stores goals which can be inferred. It is therefore not possible to use its *library(tabling)* to both store failing goals as well as retain Prolog's backtracking semantics, since `call(false)` will prevent the addition of any tabled predicate.

4   Test Framework Using Quasi–Quotations and TAP

library(xsd) has been developed in a test–driven approach. Currently its compliance to the XSD standard is ensured by more than 350 tests. Their definitions and the provided test framework take more than three times the lines of code as the core library. It has been used in a continuous integration environment using the *Travis CI*[9] service.

We have implemented a test framework based on the *Test Anything Protocol* (TAP) [6]. The SWI–Prolog package *library(tap)*[10] generates a TAP–conform text output. This interface is supported by a wide range of tools for running, rendering and analysing the test results.

The test framework is based on normal XSD and XML documents. XSD documents can be directly used. Since a single XSD should test only a small, specific

---

[8] *library(xsd)* provides the options `'without-tabling'(Bool)` and `profile(Bool)`. The example in `/test/example/choice_minmax` returns: without memoisation 0.55s with 3,628,657 inferences; with memoisation 0.01s with 50,370 inferences (SWI–Prolog 7.7.5).

[9] https://travis-ci.org/ [accessed 12 March 2018].

[10] https://github.com/mndrix/tap [accessed 12 March 2018], The Unlicense.

aspect of the validator, it is possible to define various test cases for each XSD, e.g., satisfactory and failing documents. To place all XML test documents in a single Prolog file, we have used quasi–quotations [7]. They had been added to SWI–Prolog in version 6.3.17 and are a good mean to embed external domain–specific languages into SWI–Prolog without any modification [20,21]. This way the XML can be easily annotated directly from within Prolog. The example XML of Fig. 3, which should be recognised as non–valid, is embedded into the Prolog source code of our test framework using the following snippet:

```
'missing email node'(fail): {| xml ||
  <person no="abc">
   <name>John Doe</name>
  </person> |}
```

5   Conclusion and Future Work

In this work, we have presented a declarative approach for XSD validation in SWI–Prolog. Due to its backtracking and unification mechanisms, Prolog suits very well for implementing an XSD validator which processes the XSD and XML document simultaneously in a top–down manner. We have presented an alternative XML representation in Prolog. It is based on three dynamic predicates which are asserted for the given nodes. This flattening results in a non–nested representation that can be easily queried. With the help of the presented inductive rule to generate new unique identifiers, it is simple to find all ancestors, descendants and siblings of a given XML node using unification.

The *library(xsd)* is available at https://github.com/jonakalkus/xsd and published under MIT License. It requires SWI–Prolog of at least version 7.3.26. Because it has been developed in a test–driven approach, it provides a decent test framework with currently more than 350 tests. Although not yet feature–complete, this covers the bigger part of XML Schema 1.0.[11]

To support features of the not yet widely adopted XSD 1.1 standard, *library(xsd)* currently misses support for XPATH expressions. Although SWI–Prolog provides a *library(xpath)*, it is not compatible with our flattened representation of the XML documents.

The current implementation only uses the memoisation technique presented in Sect. 3 as a first optimisation. Although the presented approach with Prolog's backtracking is very flexible, it is not optimal: given an XSD which only defines an unbounded sequence of repeating elements, we would expect that its execution time is linear to number of XML nodes. But due to the backtracking in finding both the possible elements as well as possible schema rules, this is not the case as of yet.[12] To achieve a better performance, the given XSD file should be analysed

---

[11] A list of currently supported XSD features can be found at https://github.com/ jonakalkus/xsd/blob/master/FEATURES.md [accessed 12 March 2018].

[12] The example in **/test/example/sequence_unbounded*** returns: XSD with 200 lines of code in 0.11s with 281,288 inferences; 400 lines of code in 0.71s with 1,042,288 inferences (SWI–Prolog 7.7.5).

in advance to create an individual validator. This generated Prolog program can then be used to validate XML files according to the XSD.

References

1. Bray, T., Paoli, J., Sperberg-McQueen, C.M., Maler, E., Yergeau, F.: Extensible markup language (XML). World Wide Web J. **2**(4), 27–66 (1997)
2. Fallside, D.C., Walmsley, P.: XML schema part 0: primer second edition. W3C Recommendation (2004)
3. Gao, S., Sperberg-McQueen, C.M., Thompson, H.S., Mendelsohn, N., Beech, D., Maloney, M.: W3C XML schema definition language (XSD) 1.1 part 1: structures. W3C Candidate Recommendation (2009)
4. Clark, J., DeRose, S., et al.: XML path language (XPath) version 1.0 (1999)
5. Wielemaker, J., Schrijvers, T., Triska, M., Lager, T.: SWI-Prolog. Theor. Pract. Log. Program. **12**(1–2), 67–96 (2012)
6. Specification of the Test Anything Protocol. https://testanything.org/tap-specification.html. Accessed 12 Mar 2018
7. Wielemaker, J., Hendricks, M.: Why it's nice to be quoted: quasiquoting for prolog. In: Proceedings of 23rd Workshop on Logic-based Methods in Programming Environments (WLPE) (2013)
8. Beckett, D., McBride, B.: RDF/XML syntax specification (revised). W3C recommendation 10(2.3) (2004)
9. Wielemaker, J., Schreiber, G., Wielinga, B.: Prolog-based infrastructure for RDF: scalability and performance. In: Fensel, D., Sycara, K., Mylopoulos, J. (eds.) ISWC 2003. LNCS, vol. 2870, pp. 644–658. Springer, Heidelberg (2003). https://doi.org/10.1007/978-3-540-39718-2_41
10. Wielemaker, J., Huang, Z., Van Der Meij, L.: SWI-Prolog and the web. Theor. Pract. Log. Program. **8**(3), 363–392 (2008)
11. Wielemaker, J., Beek, W., Hildebrand, M., van Ossenbruggen, J.: Cliopatria: a SWI-Prolog infrastructure for the semantic web. Semant. Web **7**(5), 529–541 (2016)
12. Goldfarb, C.F., Rubinsky, Y.: The SGML Handbook. Oxford University Press, Oxford (1990)
13. Wielemaker, J.: SWI-Prolog SGML/XML parser. SWI, University of Amsterdam, Roetersstraat 15, 1018 (2005)
14. Seipel, D.: Processing XML-documents in prolog. In: Workshop on Logic Programming (WLP) (2002)
15. Seipel, D., Baumeister, J., Hopfner, M.: Declaratively querying and visualizing knowledge bases in XML. In: Seipel, D., Hanus, M., Geske, U., Bartenstein, O. (eds.) INAP/WLP-2004. LNCS (LNAI), vol. 3392, pp. 16–31. Springer, Heidelberg (2005). https://doi.org/10.1007/11415763_2
16. Nogatz, F., Frühwirth, T.: From XML schema to JSON schema: translation with CHR. In: Proceedings of the 11th International Workshop on Constraint Handling Rules (2014)
17. Frühwirth, T.: Theory and practice of constraint handling rules. J. Log. Program. **37**(1), 95–138 (1998)
18. Swift, T., Warren, D.S.: XSB: extending prolog with tabled logic programming. Theor. Pract. Log. Program. **12**(1–2), 157–187 (2012)
19. Desouter, B., Van Dooren, M., Schrijvers, T.: Tabling as a library with delimited control. Theor. Pract. Log. Program. **15**(4–5), 419–433 (2015)

20. Nogatz, F., Seipel, D.: Implementing GraphQL as a query language for deductive databases in SWI-Prolog using DCGs, quasi quotations, and dicts. In: Proceedings 30th Workshop on Logic Programming (WLP) (2016)
21. Seipel, D., Nogatz, F., Abreu, S.: Domain-specific languages in prolog for declarative expert knowledge in rules and ontologies. Comput. Lang. Syst. Struct. **51**, 102–117 (2018). https://doi.org/10.1016/j.cl.2017.06.006

plspec – A Specification Language
for Prolog Data

Philipp Körner[✉][iD] and Sebastian Krings[iD]

Institut für Informatik, Universität Düsseldorf, Universitätsstr. 1,
40225 Düsseldorf, Germany
p.koerner@uni-duesseldorf.de, krings@cs.uni-duesseldorf.de

Abstract. In general, even though Prolog is a dynamically typed language, predicates may not be called with arbitrarily typed arguments. Assumptions regarding type or mode are often made implicitly, without being directly represented in the source code. This complicates identifying the types or data structures anticipated by predicates. In consequence, Covington et al. proposed that Prolog developers should implement their own runtime type checking system.

In this paper, we present a re-usable Prolog library named *plspec*. It offers a simple and easily extensible DSL used to specify type and structure of input and output arguments. Additionally, an elegant insertion of multiple kinds of runtime checks was made possible by using Prolog language features such as co-routining and term expansion. Furthermore, we will discuss performance impacts and possible future usages.

Keywords: Prolog · Runtime checks · Type system
Data specification

1 Introduction

In general, even though Prolog is a dynamically typed language, predicates may not be called with arbitrarily typed arguments. Assumptions regarding type or mode are often made implicitly, without being directly represented in the source code. In general, calling a predicate with an unintended argument might lead to stack overflows, infinite loops or any kind of undesired behavior. This complicates identifying the types or data structures anticipated by predicates.

For instance, assume you want to call a Prolog predicate in a newly acquired library. Documentation reveals that it implements the desired functionality, yet the call fails. The cause is ambiguous: it could be that the input was as intended, but no solution exists. Another possibility is that the input is unintended, but a call to a transformation predicate beforehand would have solved the issue.

Ideally, available documentation can be used to resolve any ambiguities. However, documentation in natural language has its limits: it cannot convey the entirety of information precisely and often gets outdated when changes are

© Springer Nature Switzerland AG 2018
D. Seipel et al. (Eds.): DECLARE 2017, LNAI 10997, pp. 198–213, 2018.
https://doi.org/10.1007/978-3-030-00801-7_13

made to the code. As an example, consider the following excerpt taken from the documentation of `member/2` as implemented in SWI-Prolog [21]:

"member(?Element, ?List) is true if Element occurs in the List."

One issue is that behavior is entirely undefined in case the second argument is not a list. In consequence, one cannot distinguish between failures such as `member(a, [b,c,d])`, where the second argument is a list but does not contain the element `a`, and `member(a, a)`, where the second argument is not a list.

In its current implementation, the predicate succeeds even if the second argument is not a proper list, i. e., a list not terminated by `[]`. In consequence, a call such as `member(a, [a,b|x])` is successful. Judging by the documentation alone, it remains unclear whether this is intended.

To overcome the limitations of documentation and to gain automatic verification, Covington et al. proposed that Prolog programmers should implement their own ad-hoc runtime type system [3]. Instead, we argue that by making use of Prolog language features, a simple and easily extensible DSL can be shipped as a reusable library called *plspec*.

The library is open source and freely available under MIT license. It can be downloaded from the GitHub Repository found at https://github.com/wysiib/plspec. It has been tested with both SWI Prolog and SICStus Prolog.

plspec is heavily influenced by *clojure.spec* [5], which was recently added to Clojure. The motivation for *clojure.spec* is similar to the one for *plspec*. Both languages are dynamically typed, often rendering it hard to identify which data should be passed to functions and what values are returned. Additionally, nested data structures can be large and confusing to inspect without tool support. Both libraries enable describing data based on construction out of small and simple building blocks. *clojure.spec* utilizes functions as building blocks, while *plspec* maintains a database of specifications described by Prolog terms.

In Prolog, we can insert runtime checks in order to distinguish between failures due to the absence of solutions and failures caused by malformed input data. Furthermore, we can check whether variables are bound to invalid values inside of the called predicate. These kinds of errors might be hidden if the predicate fails later on due to unrelated reasons. Finally, we can add guarantees that if a predicate was called in a certain way and succeeds, variables will be bound to data in a specific format.

Note that *plspec* is more than a simple type checker for Prolog's type system. Rather, it can be seen as an additional optional [2] dependent type system:

- *plspec* does not change the semantics of annotated Prolog programs, as long as specifications are implemented correctly and the program adheres to them. In case specifications are violated, an error handler is called in addition.
- *plspec*'s annotations are entirely optional. In particular, one can only partially annotate predicates.
- Specs may be instrumented in order to take into account runtime values. In this case, *plspec* specifications define a system of dependent types.

In the following, we will focus on how *plspec*'s annotations can be instrumented for different types of runtime checks, including traditional contracts [12] by specifying pre- and postconditions as well as invariants on variables.

2   Usage and Semantics

Our goal is to associate predicates with information regarding type, form and mode of arguments, most importantly what a valid argument looks like.

In order to describe data, we use so-called *specs*. A spec is either defined by a programmer by registering it via an interface predicate, a combination of multiple existing specs or one of following built-ins.

2.1   Built-in Specs

We implemented most predicates that can be used to examine terms as atomic specs. These are `float`, `integer`, `number`, `atomic`, `atom`, `var`, `nonvar` and `ground`. To verify that a term matches its spec, we call the built-in Prolog predicates with the same name, ensuring that these specs bear the common meaning and are easy to understand. Additionally, we add `any` to describe any Prolog term.

Furthermore, one can describe non-scalar data using recursive specs. The spec `list(X)` is matched if and only if the value is a (potentially empty) list of elements satisfying the spec `X`. Lists with a fixed length can be described via `tuple(X)`, where `X` is a list of specs which describe the element in that position. As an example, `tuple([integer, atom])` is matched by the value `[3, a]`, but neither `[a, 3]` nor `[3, a, b]`.

Compound terms can be described via `compound(X)`, where `X` is a compound term with the functor the term shall have. Its arguments have to be specs that describe what kind of data should be contained in that position of the term. For example, `compound(foo(atom, var))` is matched by `foo(bar, X)`.

Finally, specs can be combined with so-called connectives. So far, built-ins are `and(X)` and `one_of(X)`, where `X` is a list of specs. In the case of `and`, all specs have to be matched. For `one_of`, it is sufficient if at least one spec is fulfilled.

2.2   Instrumentation

Currently, *plspec* allows instrumentation of specs in three ways:

- Preconditions ensure that upon entry of a predicate, a given spec is matched.
- Invariants ensure that at all times during execution, a spec is matched or can still be fulfilled in case the argument is not ground.
- Postconditions ensure that if a predicate was called in a certain way, upon successful exit a second spec is matched.

Preconditions are a way to overcome the problems presented in Sect. 1. The idea is that all valid combinations of arguments to a predicate should be enumerated by the developer. In Prolog, there are multiple ways to call a predicate regarding instantiation of variables. However, with preconditions the developer can clearly state which calls were considered during implementation and testing.

In consequence, when using specs we can be sure that a failure of a predicate with a fulfilled precondition is intended behavior and, analogously, if the precondition is violated it is a type error.

In order to define a precondition, the interface predicate spec_pre/2 is used. Apart from the predicate, it takes a list of specs as an argument which can be understood as the argument vector passed to the predicate. It is allowed to specify multiple preconditions with the semantics that at least one precondition has to be matched. Otherwise, the error handler is called. For preconditions, the value a predicate is called with is passed to the predicate implementing the spec.

```
:- plspec:spec_pre(even_pred/1, [integer]). % the precondition
:- enable_spec_check(even_pred/1).           % instrumenting it
                                             % for runtime checks
even_pred(X) :-
    0 is X mod 2.

?- even_pred(0).
true   % intended success

?- even_pred(1).
false % intended failure

?- even_pred(_).
! plspec: no precondition was matched in even_pred/1
! plspec: specified preconditions were: [[integer]]
! plspec: however, none of these is matched by: [_G1322]
ERROR: Unhandled exception: plspec_error
```

Fig. 1. An example for preconditions

An example is shown in Fig. 1. We define a predicate even_pred/1 that succeeds if the parameter is an even integer and fails for odd integers. In particular, the meaning of the spec is that only integer values are valid parameters. Otherwise, no guarantees are made whether there is correct behavior in this call, may it be failure or throwing an exception.

Thus, if we pass a variable to the annotated predicate, we do not get an exception from is/2 that the arguments are not sufficiently instantiated but rather a print and an exception from *plspec*. This standard error handler can be replaced by a custom one, for example one that calls trace in order to start the debugger at this particular point in the program.

Invariants have a more sophisticated semantic: intuitively, they specify the data structures a predicate should work with. As soon as variables are bound to a value, they are checked as far as possible according to the spec. If the binding involves other variables, their check will be delayed until they get bound.

When a variable is bound to anything that cannot satisfy the spec anymore, the error handler will be called. One can specify invariants via `spec_invariant/2`. Again, the second argument is a list of specs with the same interpretation as above, i.e., for invariants the spec predicate is only called with ground values.

```
:- plspec:spec_invariant(invariant_violator/1, [atomic]).
:- enable_spec_check(invariant_violator/1).
invariant_violator(X) :-
    X = [1], X == [2]. % fail in a sophisticated way
invariant_violator(a).

?- invariant_violator(a).
true.

?- invariant_violator(_).
! plspec: an invariant was violated in invariant_violator/1
! plspec: the spec was: atomic
! plspec: however, the value was bound to: [1]
ERROR: Unhandled exception: plspec_error
```

Fig. 2. An example for invariant violations

This allows uncovering the kind of programming error shown in Fig. 2: there, we call the predicate `invariant_violator` with an anonymous variable. In the first rule, it will be bound to the list [1]. However, the specification of the argument to `invariant_violator` says that it should be atomic if bound. Since [1] is neither a variable nor atomic, the error handler is called.

If we would not specify this invariant, the first rule would fail since [1] is not equal to [2]. Thus, Prolog would backtrack into the second rule and bind the variable to the atomic value a. The invalid binding of X to [1] could not be determined without reading the source code. In particular, unit tests could never expose this issue. This kind of programming errors might trigger unintended co-routines whose effects might be hard to pinpoint.

Invariants are implemented by making use of co-routines. Thus, if the Prolog implementation does not support this feature, only pre- and postconditions are available. If the application itself uses co-routines, the effect depends on the execution order. However, as long as these co-routines do not fail beforehand, it has no influence on *plspec*.

Postconditions specify that if a certain condition held upon entry of a predicate, a second condition is implied on success. As for preconditions, the resulting value is used in order to call the predicate implementing the spec.

In particular, this allows to specify a promise that variables will be bound to values of a specified type. No promise is made if the predicate fails since no variables are bound then.

In *plspec*, one can use one or more instances of `spec_post/3` for postconditions. Apart from the predicate, it takes two lists of specs understood as argument vectors. The semantics is that if the first list of specs matches when the predicate is called, the second list of specs has to match if the predicate succeeds.

```
:- spec_post(my_member/2, [any, var], [any, list(any)]).
:- spec_post(my_member/2, [var, list(int)], [int, list(int)]).
my_member(H, [H|_]).
my_member(E, [_|T]) :-
    my_member(E, T).
```

Fig. 3. An example for postconditions

```
:- defspec(tree(X), one_of([compound(node(tree(X),X,tree(X))),
                            atom(empty)])).
```

Fig. 4. A Spec for a tree of a given type

In Fig. 3, we define two postconditions for an implementation of the member predicate. The first postcondition guarantees that if the predicate succeeds and the first argument was a variable, then it will be bound to a list. A different promise is made in the second precondition: if now the first parameter of the call is a homogeneous list of type int, the second one is a variable and the predicate succeeds, then the variable will be bound to a value of type int. In case not postcondition matches, e.g., for my_member(1,[1,2]), nothing is checked.

3   Implementation

Specifications which are readable and easy to understand are useful for documentation purposes without any additional code being executed. In this section, we will explain how we maintain the spec database, how specs are validated and how we instrument the annotations described in Sect. 2 for runtime checks.

3.1   Maintenance and Addition of Specs

Specs are stored in Prolog's fact database. For simplicity, we distinguish between different kinds of specs that are handled separately. The reason for this is that they have different roles. Since *plspec* was designed with extensibility in mind, users can define specs themselves and add them to *plspec* dynamically.

In the following, we present the reason for distinguishing between different kinds of specs and present each of them. Built-in specs are implemented in the same way users could implement them without modifying *plspec*'s source code.

Aliasing. defspec/2 allows defining new specs via composing existing ones. The first argument is an alias for the resulting spec, while the second argument consists of other specs. Recursive specs are allowed. However, they should consume at least one bit of information of a term in order to avoid infinite loops.

A built-in alias for integer is int. In the database, they are stored as a dynamic fact that maps the alias to the composition of specs. If an alias is encountered by the verification predicate, it just looks up its definition and continues with that spec.

Newly defined specs might also be compound terms which pass information, e. g., inner specs, to the other specs in form of variables. As an example, Fig. 4 shows how to define a spec for a tree of elements of a given type. A tree is defined to be either the atom `empty` or a compound term with the functor `node` and three arguments: the first and last argument are trees of the same type, whereas the middle argument is any value of the given type.

Valid values for `tree(int)`, a tree of integers, include `node(empty,1,empty)` and `empty`. Neither `node(empty,not_an_integer,empty)`, where the middle value is not of the given type, nor `tree(empty,1,empty)`, where the functor does not match, are valid.

Verification via Predicates. Another option is to implement a spec via a predicate that succeeds if a value is valid and fails otherwise. This can be achieved with `defspec_pred/2`, where the first argument is the new spec and the second is the predicate used for validation, possibly with some arguments specified.

Again, new specs might be compound terms and pass information to the predicate. The value that should be checked will always be appended as last argument to the predicate call.

Note that this implementation of specs is only suitable for values that are bound in a single unification step. Otherwise, another mechanism should be used as shown below. As an example, we can reuse the predicate `even_pred/1` from Fig. 1 which tests whether an integer is even or not. In order to use this predicate as a spec, it can be defined by `:- defspec_pred(even, even_pred)`.

Then, every time the spec `even` is used, `even_pred/1` is called with the value as argument. If it fails, the value is considered invalid. Since `even_pred/1` was annotated earlier, it will throw an exception if the value is not an integer.

Regarding built-ins, most atomic specs like `integer` or `nonvar` are implemented this way. When such a spec is encountered in *plspec*, the predicate is simply called with the current value.

Thus, this predicate should not have any side-effects or bind variables used in the passed term which might fire additional co-routines. In fact, checking specifications at runtime should not interfere with the execution of the annotated program in any way. In order to ensure this, we copy each term before using it to check a *plspec* annotation. If the spec predicate succeeds, the original term is compared to its copy. If a variable was bound, an error message will be printed.

Recursive Spec Predicates. The third way to define specs is more involved. If a value is not bound in a single unification step but rather "consumes" only some part of the value, an appropriate spec can be registered by calling `defspec_pred_recursive/4`.

Recursive specs can be implemented based on a predicate verifying a part of the property, the "consumption" mentioned above. Afterwards, it hands back control to *plspec* and exposes new specs and variables that should be checked.

This predicate is the second argument to `defspec_pred_recursive/4`. It will be called with all arguments directly wired in the spec definition. Additionally, the value is passed to the predicate. The last two arguments to that predicate are two variables. The first variable should be bound to a list of specs and the second variable to a list of values which might still be variables themselves. *plspec* will take these values and check them against the returned specs.

The third argument to `defspec_pred_recursive/4` is a predicate which merges the results of those checks. The basic operations and as well as or already are implemented and can be used. If a property like "exactly m out of n specs shall be true" is desired, this predicate has to be implemented by the user.

Finally, the fourth and last argument is the merge predicate which is called for invariant checks. It has to account for the fact that values might not be fully instantiated yet. In *plspec*, this predicate is implemented using co-routines in order to wait for further instantiation of the data to be verified. `and_invariant` as well as `or_invariant` are already implemented.

Internally, we implemented the checks for compound terms, lists and tuples like this. The functor of a compound term is immediately checked. Following, the specs of its arguments and the current values are returned because they might involve variables that are bound later.

As an example, consider the spec `list(int)` and the value `[1,X|T]`. A given list is deconstructed as far as possible in order to check the outer spec, i. e., the value is actually is a list. Then, the inner spec `int` is repeated for all elements. Here, we check that both `1` and `X` are integers. Since `X` is a variable, this check is handled by a co-routine that fires once `X` is bound. In presence of non-instantiated tails, the outer spec is kept and delayed until further instantiation. This means, a co-routine is set up that recursively checks that `T` also matches the spec `list(int)`. The spec `tuple(_)` is implemented similarly. In both cases, the resulting specs need to be merged with and.

Connectives. Connectives are specs that do not consume any part of a value. While they are implemented exactly like the recursive specs above, they are stored separately. Many connectives might have infinite equivalent specs, e. g., `int` is the same as `or([int])` and `or([int, int])`. Thus, connectives are avoided when enumerating possible specs for a value.

These kind of specs are registered by calling `defspec_connective/4`, where arguments and semantics exactly match those of `defspec_pred_recursive/4`. As above, built-in examples are `one_of` as well as `and`, which allow specifying at least one or all specs have to match a value. `one_of` is implemented with or as the merge predicate.

3.2 Instrumenting Specifications for Runtime Checks

In order to insert runtime checks for the properties specified in *plspec* annotations, we make use of term expansion, i.e., source-to-source transformation.

Since annotations can also function as plain documentation, the user can explicitly state which predicates should be expanded by inserting runtime checks

```
1    my_member(A, B) :-
2      ([[var, list(any)], ...]=[] -> true
3      ; plspec_some(spec_matches([A, B], true), [[var, list(any)], ...])
4          -> true
5          ;   error_handler_pre(my_member/2, [A, B], [[var, list(any)], ...])),
6      ([[any, list(any)]]=[C]
7          -> lists:maplist(plspec:invariant_check(my_member/2), C, [D, [E|F]])
8          ;   true),
9      [A, B]=[D, [E|F]],
10     plspec:which_posts([[any, var]], [[any, list(any)]], [D, [E|F]], G, H),
11     my_member(D, F),
12     lists:maplist(plspec:check_posts([D, [E|F]]), G, H).
```

Fig. 5. Expanded recursive rule

utilizing the given annotations. We will explain the term expansion on the example of the second, recursive rule of our `my_member/2` predicate shown in Fig. 3.

Consider Fig. 5: in lines 2–5, we check whether any precondition is specified. If there is at least one precondition, the `plspec_some` call will check whether at least one precondition is satisfied and an error is thrown. If no precondition was satisfied, no check will be performed. The check will simple try to conform each spec with each value the predicate was called with.

Afterwards, specified invariant checks are set up in lines 6–8. Note that there is no call to an error handler yet. Instead, the check and potential error handling happens inside of co-routines which will be described in more detail later.

The unification with the head of the rule happens in line 9. Note that A and B in line 1 are fresh variables. Otherwise, if the arguments do not unify with the head, we would not have an opportunity to catch potential errors there.

In line 10, the premises of the implications stated for postconditions are verified. Conclusions of the postconditions and whether they hold are checked again in line 12. The error handling for postconditions is not shown here because it is part of the `check_posts` predicate. Between these two steps that verify the postcondition, the original goal remains in line 11. This ensures the correct values are used for both parts of the postcondition.

3.3   Co-routining for Invariants

Invariants are violated as soon as variables are bound to incorrect values. This can be checked by setting up a number of co-routines.

`defspec_pred` is a special case of `defspec_pred_recursive`: it consumes the entire value in one go without producing new values. The trade-off is that values for this kind of spec must be bound in a single step. Otherwise, the co-routine that blocks until the value is not a variable anymore fires on a partially instantiated term and fails. On the other hand, blocking until a value is ground does not catch errors where partial instantiation is undesired. This allows easy implementations because no internal structure of a term has to be exposed.

On the other hand, `defspec_pred_recursive` produces new specs and new values. For example, one can bind a variable to a compound term with a given functor but bind its arguments later on. These arguments as well as their

```
and_invariant([], [], _, true).
and_invariant([HSpec|TSpec], [HVal|TVal], Location, R) :-
    setup_check(Location, ResElement, HSpec, HVal),
    and_invariant(TSpec, TVal, Location, ResTail),
    both_eventually_true(ResElement, ResTail, R).

both_eventually_true(V1, V2, Res) :-
    when((nonvar(V1); nonvar(V2)),
         (V1 == true -> freeze(V2, Res = V2)
        ; nonvar(V1) -> Res = V1
        ; V2 == true -> freeze(V1, Res = V1)
        ; nonvar(V2) -> Res = V2)).
```

Fig. 6. An implementation of **and** based on co-routines

corresponding specs have to be exposed to *plspec*, that will set up new co-routines on them in return. This way, all invalid bindings of variables can be accounted for.

The tricky part is that results of subterms usually only propagate one at a time. If the third argument of a compound term is bound incorrectly, but the first argument remains a variable, *plspec* has to immediately fail. Otherwise, the first variable might not be bound at all and the error would go unnoticed.

Thus, a second merge predicate able to deal with co-routines is required. An implementation merging the results with the connective **and** is shown in Fig. 6.

The predicate **setup_check** will set up co-routines in the same way as the original spec did, using the exposed structure of terms. If the check succeeds, **ResElement** is bound to **true** or, otherwise, an error term containing a reason.

The connective is chained between the results. For example, if the term **foo(1, a, X)** is matched against **compound(foo(int, atom, var))**, the predicate **int(1), atom(a), var(X)** is formed. Each of the three calls is set up individually using its own co-routine. As soon as one fails, the entire formula is false and all co-routines are terminated by unifications in **both_eventually_true**.

Analogously, in order to implement **or**, a single **true** suffices in order for the formula to be true and to terminate all co-routines that were set up on the other disjuncts. Additionally, it has to be propagated when all disjuncts fail in order to throw an error. However, it is enough to check all alternatives only when we can determine *all of them*. Because we only want to raise an error if the entire disjunction evaluates to false but one alternative cannot be evaluated yet, we can understand non-termination as "still possible".

4   Performance Impact

Since specs are checked at runtime, naturally there is an overhead. In this section, we discuss which predicates should be annotated by measuring the performance impact caused by the runtime checks. As a first example, we consider **member/2** that succeeds if the second argument is a list containing the first argument.

```
member(Element, [Element|_Tail]).        member_entry(Element, List) :-
member(Element, [_Head|Tail]) :-            member(Element, List).
    member(Element, Tail).
```

Fig. 7. Definition of member/2

```
:- spec_pre(member/2, [any, one_of([var, list(any)])]).
:- spec_invariant(member/2, [any, list(any)]).
:- spec_post(member/2, [any, any], [any, list(any)]).
```

Fig. 8. Possible specs of member/2

```
:- spec_pre(reverse/3, [list(any), list(any), var]).
:- spec_pre(reverse/3, [var, list(any), list(any)]).
:- spec_invariant(reverse/3, [list(any), list(any), list(any)]).
:- spec_post(reverse/3, [list(any), list(any), var],
                        [list(any), list(any), list(any)]).
reverse(L, Rev) :-
    reverse(L, [], Rev).
reverse([], Acc, Acc).
reverse([H|T], Acc, Rev) :- !,
    reverse(T, [H|Acc], Rev).
```

Fig. 9. Annotated version of reverse

In Fig. 7, the definition of member/2 is shown. Additionally, we define a predicate member_entry/2 that wraps the member/2 predicate. One could argue, that valid calls to member/2 should have a list as a second argument. While it is totally sound that the predicate just fails if the second argument is not a list, in most cases such a call indicates a programming error somewhere in the code.

Thus, we add annotations to member/2 and, analogously, to member_entry/2 as shown in Fig. 8. The spec_pre directive allows that the element might be of any type, but the second argument is either a variable or a proper list. Secondly, spec_invariant ensures that if the second argument is bound, it still has to be possible for it to become a proper list. Lastly, spec_post guarantees that if the predicate succeeded for any input, the second argument will be a proper list.

We consider three benchmark configurations: first, the predicate is not annotated with a spec. Second, a spec is applied to the entry point, but not the recursion. Third, the spec is checked in each recursion step.

These calls are made to member/2 with an integer $Index$ and a list of integers ranging from 1 to N, and to reverse/2 with the same list and a variable. Additionally, we benchmarked calls to reverse with an accumulator that is implemented and annotated as in Fig. 9. For the entry level benchmark, we only annotate reverse/2, dropping the second spec in each of the argument vectors. Each run is repeated ten times and the median runtime is given. All benchmarks

Table 1. Runtimes and inference count of multiple kinds of annotations

Program	Index	Runtime (msecs)			Inferences		
		len $=100$	len $=1000$	len $=2500$	len $=100$	len $=1000$	len $=2500$
member	50	0	0	0	51	51	51
	100	0	0	0	101	101	101
	500	0	0	0	103	501	501
	1000	0	0	0	103	1001	1001
	2500	0	0	0	103	1003	2501
member-entry	50	0	7	18	5224	50224	125224
	100	0	7	19	5274	50274	125274
	500	0	7	20	3944	50674	125674
	1000	0	7	18	3944	51174	126174
	2500	0	5	18	3944	38144	127674
member-recur	50	37	462	1168	254317	3077617	7783117
	100	61	890	2310	352621	6010892	15441392
	500	53	3583	11447	284214	23807092	71037592
	1000	48	4939	20756	284214	31877371	126357842
	2500	52	3994	29653	284214	25339314	197818621
reverse		0	0	0	103	1003	2503
reverse-entry		1	15	34	10250	101150	252650
reverse-recur		214	23549	174418	1171905	113417205	707292705

were run on an Intel(R) Core(TM) i7 CPU running at 2.60 GHz. We used SWI
Prolog version 7.6.4 and configured it to use increased stack size by starting it
with the parameters -G200g -T40g -L4g. Benchmarks were run sequentially to
avoid issues due to scheduling or hyper-threading.

Table 1 depicts the results of the benchmarks. Columns show the length of
the list split by runtime of the query as well as amount of inferences. For the
member predicates, lookups of different indices are benchmarked in each row.
The programs "member" and "reverse" stand for the original predicates without
annotations, whereas the suffix "entry" and "recur" distinguish between the
annotation at entry-level and recursion-level respectively.

As depicted in Table 1, for both member and reverse, the amount of additional inferences and runtime is roughly constant (but large) if only the entry
level is annotated.

However, if the specs are checked in every single recursion step, for member,
the overhead quickly grows linearly in the length l of the list as well as linearly
in the index i that is looked up, causing a quadratic overhead of $i * l$.

The overhead for reverse actually grows quadratic in the size of l. This is
because in every step, the entire list without its head is validated against the spec
again. We can clearly see that this becomes very slow even if list size increases
moderately and such instrumentation should be avoided.

Since this overhead is enormous, recursive predicates should not be annotated. Instead of checking the same property again and again, one can annotate an invariant on entry level. Then, the performance impact is less noticeable.

5   Related Work

The idea of integrating runtime checks based on annotations into Prolog is not new. In [18], the authors present the library type_check that implements an optional Mycroft-O'Keefe type system [14] for SWI and YAP Prolog. In comparison to *plspec*, type_check supports type variables as well as static type checks. However, mode annotations are not enforced. Thus, it is not possible to ensure that variables are instantiated before or after a call to a predicate and the semantics for (runtime) type checking is similar to invariants in *plspec*.

On the other hand, annotating pre- and postconditions has, for instance, been suggested in [9]. In contrast to our approach, the authors extend the usual notion of pre- and postconditions by annotations attached to the Prolog ports for fail and redo. In consequence, they work closer to the execution model of the underlying Prolog interpreter. Furthermore, the author provides the calling context, e. g., the parent predicate, to the specification under test. This allows for more fine-grained reasoning. Our approach on the other hand provides checking of invariants at any point of Prolog execution by means of co-routines.

The work around assertion checking in CiaoPP [17], uses abstract interpretation to try to discharge assertions at compile time. Assertions which cannot be checked statically are performed at runtime, using program transformation. To our knowledge, CiaoPP only supports Ciao Prolog. While *plspec* requires co-routining for its full functionality, pre- and postconditions work with any Prolog implementation that supports term expansion.

A different approach to testing has been followed in [13]. In contrast to our approach, the authors do not focus on the introduction of runtime checks into Ciao Prolog, but rather try to unify unit testing and runtime checking. This way, only one kind of annotation is needed for different testing purposes. We extend upon this work by the introduction of invariance annotations and the ability to use connectives as discussed in Sect. 3.1. So far, we have not evaluated if we can extract unit tests from our annotations, but intend to do so.

Documentation of Prolog code has been considered in [20], where the authors introduce *PlDoc*, a documentation format used for literate programming. The corresponding Prolog package has since been included in SWI Prolog. Instead of integrating documentation into the Prolog code itself, the LaTeX package *pl* [15] embeds code into the documentation. Using the package, a single source file can be run both by any LaTeX binary and a Prolog interpreter.

Aside of Prolog, other declarative logic programming languages feature comparable systems. Mercury [19] includes a type system [4, 7] together with a set of mode annotations [16]. However, the type system implemented in Mercury differs from the one we suggested: Though it supports higher-order functions, it neither allows types to be defined by a predicate nor to define a union of two types. In contrast to *plspec*, Mercury allows for type variables to be used. This makes it possible to specify, for instance, that the output of a function will have the exact same type as the input, regardless of the type itself.

Similar annotations to those in *plspec* can be found in Erlang's type specification language [8]. These are used, e. g., in the program analyzer Dialyzer [11]. In Erlang, it is only possible to create new types by defining a union of two existing types, which may be pre-defined or an atomic singleton like the number 42 or the atom foo. As discussed in Sect. 3.1, *plspec* allows to define a type for all values that fulfill a given predicate.

Furthermore, Erlang allows specifying types for higher-order functions which *plspec* does not support. Function specifications in Erlang can be regarded as pre- and postconditions in *plspec*. Just like Mercury, Erlang supports type variables.

6   Future Work

While *plspec* is capable of exposing real errors in real world Prolog applications, several improvements to the library should be made:

- The default error messages have room for improvement. Whenever possible, the smallest subterm that makes a spec invalid should be included separately. This allows developers to identify faster and easier what went wrong.
- We can imagine adding further annotations. For example, it can be desired that co-routines are terminated when a certain predicate succeeds or that predicates must never fail given their precondition is fulfilled.
- In Sect. 4, we found that checking annotations of recursive predicates is very slow. If we added static analysis or used gradual typing, most of that overhead could be avoided. For example, a meta-interpreter that makes use of *plspec*'s annotations could be employed for static type checking.
- Support for type variables should be integrated into the specification language, increasing both expressiveness and value for documentation.

Apart from documentation and runtime checks, there are several applications that could benefit from these annotations and may be subject of future research.

In order to reduce the burden on programmers and increase applicability, it is desirable that for existing code, one does not have to write specs by hand. Due to the logical and declarative nature of Prolog, we can easily find matching specs to a given value by calling the verification predicate with a variable for the spec. While this allows us to generate a spec for a given value, it is not yet possible to generate a spec that matches all elements in a *series* of data.

If this functionality existed, one can think further: with additional tool support, specs as well as entire contracts could be inferred, for example, simply by running unit tests that contain only calls which are known to be valid.

Furthermore, some of these annotations could be re-usable for a partial evaluator such as LOGEN [10]. An issue with LOGEN is that even though its binding-time analysis already generates annotations, usually its user has to improve them manually. Some information that *plspec* covers, e. g., how predicates are intended to be called, might reduce the manual work required.

Another area is data generation based on a spec. We could use our annotations to generate arbitrary data featuring a certain structure or other properties.

This could be achieved by linking *plspec* to existing test frameworks for Prolog such as [1]. The authors follow an approach to test case generation and shrinking similar to Erlang's QuickCheck [6]. However, we would regard test failures as failing predicates if a spec is matched. In consequence, we would not describe actual output values in terms of input values.

Besides, often predicates only transform data into a different structure. With annotations that precisely describe different data structures passed to and returned from a predicate, it might be feasible to both repair incorrect and synthesize new programs solely based on *plspec*'s annotations.

Finally, *plspec* could make use of existing annotations, for example mode or meta-predicate annotations. They could be converted directly into our format.

7   Conclusion

In this paper, we presented the library *plspec*. It provides a DSL that can be used to document Prolog predicates in a way that is straightforward. This DSL is easily extensible without getting involved with internal implementation details and flexible enough to suit the needs of a broad range of Prolog programs. Furthermore, these annotations can be used in order to quickly and effortlessly enable runtime checks if required.

While the performance hit might be too big for recursive predicate, we argue that, firstly, most checks suffice to be made at the entry level because of the recursive implementation of specs for recursive data. Furthermore, invariants are powerful enough to catch incorrect bindings at a deeper recursion level. Secondly, it suffices to annotate interface predicates in real programs. Usually, these are not implemented recursively but call auxiliary predicates, thus avoiding unnecessary runtime checks. Lastly, *plspec* is a tool intended to catch errors during development. Our runtime checks should not be deployed as production code and if so, only very carefully.

References

1. Amaral, C., Florido, M., Santos Costa, V.: PrologCheck – property-based testing in prolog. In: Codish, M., Sumii, E. (eds.) FLOPS 2014. LNCS, vol. 8475, pp. 1–17. Springer, Cham (2014). https://doi.org/10.1007/978-3-319-07151-0_1
2. Bracha, G.: Pluggable type systems. In: OOPSLA Workshop on Revival of Dynamic Languages (2004)
3. Covington, M.A., Bagnara, R., O'Keefe, R.A., Wielemaker, J., Price, S.: Coding guidelines for prolog. Theory Practice Logic Program. 12(6), 889–927 (2012)
4. Dowd, T., Somogyi, Z., Henderson, F., Conway, T., Jeffery, D.: Run time type information in mercury. In: Nadathur, G. (ed.) PPDP 1999. LNCS, vol. 1702, pp. 224–243. Springer, Heidelberg (1999). https://doi.org/10.1007/10704567_14

5. Hickey, R.: clojure.spec - Rationale and Overview (2016). https://clojure.org/about/spec

6. Hughes, J.: QuickCheck testing for fun and profit. In: Hanus, M. (ed.) PADL 2007. LNCS, vol. 4354, pp. 1–32. Springer, Heidelberg (2006). https://doi.org/10.1007/978-3-540-69611-7_1

7. Jeffery, D.: Expressive type systems for logic programming languages. Dissertation, Department of Computer Science and Software Engineering, The University of Melbourne (2002)

8. Jimenez, M., Lindahl, T., Sagonas, K.: A language for specifying type contracts in erlang and its interaction with success typings. In: Proceedings of the 2007 SIGPLAN Workshop on ERLANG, ERLANG 2007, pp. 11–17. ACM (2007)

9. Kulaš, M.: Annotations for prolog – a concept and runtime handling. In: Bossi, A. (ed.) LOPSTR 1999. LNCS, vol. 1817, pp. 234–254. Springer, Heidelberg (2000). https://doi.org/10.1007/10720327_14

10. Leuschel, M., Craig, S.J., Bruynooghe, M., Vanhoof, W.: Specialising interpreters using offline partial deduction. In: Bruynooghe, M., Lau, K.-K. (eds.) Program Development in Computational Logic. LNCS, vol. 3049, pp. 340–375. Springer, Heidelberg (2004). https://doi.org/10.1007/978-3-540-25951-0_11

11. Lindahl, T., Sagonas, K.: Detecting software defects in telecom applications through lightweight static analysis: a war story. In: Chin, W.-N. (ed.) APLAS 2004. LNCS, vol. 3302, pp. 91–106. Springer, Heidelberg (2004). https://doi.org/10.1007/978-3-540-30477-7_7

12. Mandrioli, D., Meyer, B.: Design by contract. In: Advances in Object-Oriented Software Engineering, p. 1 (1991)

13. Mera, E., Lopez-García, P., Hermenegildo, M.: Integrating software testing and run-time checking in an assertion verification framework. In: Hill, P.M., Warren, D.S. (eds.) ICLP 2009. LNCS, vol. 5649, pp. 281–295. Springer, Heidelberg (2009). https://doi.org/10.1007/978-3-642-02846-5_25

14. Mycroft, A., O'Keefe, R.A.: A polymorphic type system for prolog. Artif. Intell. **23**(3), 295–307 (1984)

15. Neugebauer, G.: pl-Literate Programming for Prolog with $\LaTeX$, (1996). https://www.ctan.org/pkg/pl, version 3.0

16. Overton, D.: Precise and expressive mode systems for typed logic programming languages. Dissertation, Department of Computer Science and Software Engineering, The University of Melbourne (2003)

17. Puebla, G., Bueno, F., Hermenegildo, M.: Combined static and dynamic assertion-based debugging of constraint logic programs. In: Bossi, A. (ed.) LOPSTR 1999. LNCS, vol. 1817, pp. 273–292. Springer, Heidelberg (2000). https://doi.org/10.1007/10720327_16

18. Schrijvers, T., Santos Costa, V., Wielemaker, J., Demoen, B.: Towards typed prolog. In: Garcia de la Banda, M., Pontelli, E. (eds.) ICLP 2008. LNCS, vol. 5366, pp. 693–697. Springer, Heidelberg (2008). https://doi.org/10.1007/978-3-540-89982-2_59

19. Somogyi, Z., Henderson, F.J., Conway, T.C.: Mercury, an efficient purely declarative logic programming language. In: Proceedings ASCS, pp. 499–512 (1995)

20. Wielemaker, J., Anjewierden, A.: PlDoc: Wiki style Literate Programming for Prolog. CoRR, abs/0711.0618 (2007)

21. Wielemaker, J., Schrijvers, T., Triska, M., Lager, T.: SWI-prolog. Theory Practice Logic Program. **12**(1–2), 67–96 (2012)

Author Index

Printed in the United States
By Bookmasters